Nor Shall My Sword

Nor Shall My Sword

*Discourses on Pluralism, Compassion
and Social Hope*

F. R. Leavis

BARNES & NOBLE BOOKS · NEW YORK
(a division of Harper & Row Publishers, Inc.)

Published in the U.S.A. 1972 by
HARPER & ROW PUBLISHERS, INC.
BARNES & NOBLE IMPORT DIVISION

ISBN 06 4941221

Printed in Great Britain

*To the York students
who gave me a new Blake with
clean margins to write in*

Contents

All thought is incarnate; it lives by the body and by the favour of society. But it is not *thought* unless it strives for truth, a striving which leaves it free to act on its own responsibility, with universal intent.

At whatever level we consider a living being, the centre of its individuality is real. For it is always something we ascertain by comprehending the coherence of largely unspecifiable particulars, and which we yet expect to reveal itself further by an indeterminate range of future manifestations. Thus the criteria of reality are fulfilled.

MICHAEL POLANYI

The vision of the machine so perfected and controlled that it would leave us free to get on with the real business of living is a potent one; it recurs throughout the nineteenth century and it remains the ideal for an automated society. Fundamental to Morris's Utopia, however, was the conviction that man, so liberated, would learn to love and create art: 'People living under the conditions of life and labour above mentioned,' wrote Morris in *A Factory as it Might be*, 'having manual skill, technical and general education, and leisure to use these advantages, are quite sure to develop a love of art . . .'

GILLIAN NAYLOR, The Arts and Crafts Movement

'I never forget when William Morris died. A handful of us were passing the *Western Daily Mercury* office, Plymouth, to attend a socialist branch meeting, and the news just choked us. The announcement was made that "Mr William Morris died today". "Who the hell's he?" said a worker to another. Just after we heard a deafening roar. The Bashites had won!'

E. P. THOMPSON, *William Morris, Romantic to Revolutionary*
(quoted by Gillian Naylor)

I

Introductory :
'*Life*' Is *A Necessary Word*

INTRODUCTORY:
'LIFE' *IS* A NECESSARY WORD

No one will suppose me to have forgotten how Blake's stanza continues: he declares that *his* sword shall not sleep in his hand

Till we have built Jerusalem

A Blake in our time (if we can imagine one) couldn't have written that—written it with the irresistible naturalness that belongs to it in its whole living context. The genius born in the middle of the 18th century had to work out his own mode of formulating (that is, determining) for himself, and communicating, his intransigent certainties. What, in his mythology, Jerusalem represents, while it certainly plays an important rôle in relation to his certainties, doesn't seem to be one of them. It is a posited goal or τέλος—one that Blake constantly fails to make anything but posited. To say 'constantly' is to recognize Blake's awareness of his failure—failure by the criteria that as an artist he presents us with. The awareness is implicit in the way in which, over the years, he tried again and again in his poetico-mythopoeic explorations to arrive at a convincingly 'created' suggestion of what would succeed the reversal of the Fall.

It is plain to me that in the nature of things the failure was inevitable; there couldn't have been anything else. That, I think, is what Lawrence meant when he said: 'Blake was one of those ghastly obscene knowers.' But—nor did Lawrence suppose so—that doesn't dispose of Blake. In fact, the way in which he compels us to recognize the truth, and the bearing of the truth, that the Laurentian ejaculation points to is a mark of the importance to us of the Blakean genius in face of the present human world.

For while, in the given regard, it is difficult to believe that Blake ever rested for long in flattering self-deception, it was not for nothing that he died singing: he couldn't but know that his life had been a triumph and vindication of human creativity. Though he was denied the public that every artist needs, he had robustness enough of creative conviction to carry him unsubdued through life, and leave him at the end fulfilled and justified. The disinterestedness that Dickens was to embody in Daniel Doyce was invincible in Blake: he was incapable of supposing that in his creativity he 'belonged to himself'. What Blake represents is the new sense of human responsibility that we may reasonably see as the momentous gain accruing to the heritage—to be taken up (that is) in the creative continuity—from among the diverse manifestations of profound change that are brought together under 'Romanticism'. The Blakean sense of human responsibility is as much the antithesis of the defiant Byronic hubris as it is of the hubris of technologico-positivist enlightenment. It goes with a realization that without creativity there is no apprehension of the real, but that if experience is necessarily creative, the creativity—as every great artist testifies —is not arbitrary; it is a Doyce-like self-dedication to a reality that we have to discover, knowing that discovery will at best be qualified by misapprehension and certainly incomplete. Our reality couldn't be 'there' for us without collaborative creation, but it is not an achievement to be credited to creative human selfhood (an essential contradiction)—or to coagulated egos.

Blake's battle against the ethos he identifies with the linked names of Newton and Locke was an insistence on human responsibility.

Save his own soul's light overhead
None leads him, and none ever led

—when Swinburne, exalting Man, wrote this he was taken as a voice of Victorian humanist enlightenment, which, like what in general is in our time called 'humanism', may fairly be called hubristic; but those very characteristic utterances of Blake's which might seem to be obviously in resonance were essentially, however they may have startled Crabb Robinson, not expressions of humanist hubris. It is not for nothing that a theologian (J. G. Davies in *The Theology of William Blake*), after a close inquiry, concludes (remarking that this will be disappointing 'to those who have glorified Blake as the great heresiarch'): 'we are bound to affirm that his doctrines fall within the general tradition of Christianity'. The same theologian writes: 'Blake was a genius and he knew it; but he also knew that this was not a cause for pride; what he was, what he did, was really the work of a higher power operating all his artistic creation through him.'

Such testimony, with its distinctive emphasis, is worth citing because of the need to dwell on what it is that makes Blake so important. The sense of human responsibility as he represents it is inimical to what is commonly meant by 'humanism' in our technologico-Benthamite world. Genius for him is a peculiar intensity and strength of representativeness: the artist's developed, conscious and skilled creativity is continuous with the creativity inseparable, he insists, from perception itself, and from all human experience and knowledge. Of its nature it can't be wilful and gratuitous—that is, irresponsible; though it may (or will) in this or that way fall short, achieving at best a limited rightness, it is always concerned for the real. To be creative in the artist's way is implicitly to assert responsibility, and Blake's distinction is to be fully conscious of that—to recognize unhesitatingly the nature of the responsibility resting on man; resting on himself.

Yet, while he was so far clear in his certainties, and though he was incapable of hubris, his thought ran out into a region where thought and imagination must necessarily incur defeat; he undertook what couldn't be done, and, in attempting it, was lost in the wordy generality that covers contradiction and confusion. The failure he implicitly recognized, but he never arrived at a perception of the inherent impossibility—how could he have done? What he did achieve justifies us in imputing to him astonishing genius. It puts us in a position to see and to say what, as achievement, it was—which is to realize its bearing on the sickness of our world.

The genius manifests itself in a profound communicated insight into the nature of human life, the human situation, and human potentiality. Blake's extraordinarily subtle presentment of the Urizen in man is in its positive significance an insistence on the creativeness that is responsibility—on the responsibility that is creative. For Urizen, who in *Hard Times* appears as Gradgrind, the important thing is that life shall be known, its possibilities determined, so that the human individual as selfhood—which is what is given us in Urizen—can feel protected by known law (in both senses of the word) from the new, unprecedented and unplaced. 'Louisa, never wonder!' says Mr Gradgrind sternly to his deplorably youthful daughter. Responsibility in this sense means scrupulous deference towards the laws that limit possibility, and towards the formulated definitions that chart the actual and may be taken as the real reality.

What 'wonder' means is what is given us in that creative lapse, or escape, from Eliotic habit, the unique and lovely 'Marina', which opens:

> *What seas what shores what grey rocks and what islands*
> *What water lapping the bow*

and, but for the brief recall of the opening that forms the three-line coda, ends

> *The awakened, lips parted, the hope, the new ships.*

Wonder is the welcoming apprehension of the new, the anti-Urizenic recognition of the divined possibility. It is the living response to life, the creative—the life of which the artist in his creativity is conscious of being a servant. To be spontaneous, and in its spontaneity creative, is of the essence of life, which manifests itself in newness that can't be exhaustively reduced to the determined, whatever some biologists may still hope. With a great diversity of creative resource Blake insists on that truth, of which he has a profound and subtle grasp, intellectual and imaginative. But in his life-long war, spiritual, poetic and mythopoeic, against the 'positive culture'[1] of the age an essential dependence on Christian tradition was inevitable; or, at least, so it must seem to us. By way of explaining the necessity of the creative battle he preached and practised, he invoked the theme of the Fall. His interpretations (they were plural) were very much his, but couldn't save him

---

[1] 'But so positive was the culture of that age, that . . . it crushed a number of smaller men who felt differently but did not care to face the fact. . . .'—T. S. Eliot.

from the impasse (for thought and creation) he never recognized as that.

The significance of the postulated Fall is that, in offering to explain the apprehension and the nisus, it gives authority to the human sense of potentiality asking to be realized that has played so important a part in cultural history. It gives authority; but we have to note that for Blake too the postulate means that the harmony lost in the Fall is what is to be recovered, it being that which gives us the τέλος of creative effort—the goal his sick hunger for which Eliot conveys so characteristically; as (for instance) in his evocation of endless time in the stanza passage of the second movement of 'The Dry Salvages'. That anguish of meaninglessness, like the life-long brooding on death and decay, goes in Eliot—and goes significantly—with his paradoxical unbelief in human creativity. I make this point because Blake's own creativity, the manifestation of his intense positiveness of apprehension and realization, failed when it came to the evoking of the restored Eternal Man, the upshot or sequela of the Apocalypse. How could there *but* be failure? The positiveness of apprehension was in the nature of things impossible there.

'To generalize is to be an idiot'; 'Truth only exists in minutely organized particulars'; 'The Infinite alone resides in Definite and Determinate Identity': here we have Blake's criteria as a poet—an intellectual poet, or poet-intellectual. His strength as an artist (to use his own comprehensive term) was to have known so effectively what he meant by them, and to have been governed by them in his practice. His strength as an artist, it may be added, was his strength as a thinker—a proposition that may be inverted. When he used the verb 'generalize' in that

pejorative way he was thinking of the problem facing anyone who aspires to present with cogent finality the essential
truths about human nature. 'Life' is a necessary word, but
what it denotes is 'there' only in the individual. Psychology
is individual psychology and is still that in its dealings with
individuals in mutual relation. When, in spite of the scientific pretension, I say 'psychology', no other word offering,
I mean psychology as a gifted novelist is concerned with
it. A great novelist's interest in the individual focus en-
gages a profound and conscious exploratory preoccupation
with what the indispensable word 'life' portends. I am
thinking of the kind of greatness Lawrence denies Forster
when he writes: 'Life is more interesting in its under-
currents than in its obvious, and E.M. does see people,
people and nothing but people ad nauseam.' Blake's genius
was to be, in the sense in which a great novelist necessarily
is, a profound psychologist. That is apparent not only in
the lyrics, but, for all the confusion and the elements of
unsuccess from which none of them is free, in the longer
poems. He presents with clairvoyant penetration and
compelling actuality the state—or rather the interacting
energies, the disharmonies, the conflicts and the trans-
mutations—of humanity as it is. When, however, he is
faced with presenting the restored condition of the Eternal
Man, the joyous resolution, he is at a loss. How can Man
be brought before us unless as *a* man? It is true that the
equivocal freedom from schematic, or logical, rigidity
with which Blake uses personification and symbolism in
general—a freedom that we find from time to time to be
strikingly justified—enables him, in dealing with the condition of Eternity, to suggest that it is compatible with
a familiar kind of human interplay between separate centres of sentience, so that the Eternal Man, the realized

inclusive total harmony and unity, can be social life too. But it is only by being vague, general and boring that Blake escapes being arrestingly paradoxical. Not merely is there, with the absence of disharmony, victim, and cross-purposes —or even duality of purpose, a vagueness about the range and reality of human motivation. Love, we are to believe, reigns in Eternity; but what can we make of the idea of love—human love—as belonging to a supremely human order (the Blakean eternal reality is essentially that) in which Man (or a man) is Woman (or a woman) too, and Woman Man, the sexes having ceased to exist? We can't but feel that the fact of jealousy, the theme that Blake, so great a master of it when dealing with the fallen world, makes so formidable, has been merely left behind—and, in a way that doesn't help us. And actually the reversal of the Fall is no more *explained* by Blake's mythopoeia than the Fall itself was. For Blake, the product of a Christian culture, it presented itself as the obvious form of explanation, and it was assumed as such, being a matter of implicit acceptance.

But if one said 'of faith', it would be only to add that 'faith' here isn't vision. 'Vision' itself in Blake exegesis has been made an insidiously equivocal word, so I will shift to 'imagination'. The Eternal Man and Jerusalem can't even by Blake be *imagined*; there can be no presentation of them in terms of 'minute particulars'. Insistent prophetic vehemence, the thickening of symbolisms, can only give us the antithesis of the 'wiry bounding line'.

I am not dismissing Blake. My aim, on the contrary, is to focus the attention on what makes him so important to us, and, what that entails, on that implication of his insight on which he himself—inevitably, it seems reasonable to say—failed to close his grasp. He could represent Los, human creativity in the fallen human condition, as working creatively though unpossessed of the vision of any justifying ultimate end, but, as the poet, he couldn't, with the Bible, Swedenborg, Boehme and Milton behind him, help feeling that he must himself aspire to a certitude and clarity of such vision. This is the aspect of Blake to which Lawrence's sharp comment applies: Blake is committed to knowing where knowing is impossible. But the comment doesn't entail a comprehensive adverse judgment. Lawrence might have said of his own works what Blake said of his paintings and designs: 'though I call them mine, I know that they are not mine'.

It is the Blake corroborated and reinforced by Lawrence that I have in mind when I contend that what desperately needs to be emphasized in the present plight of mankind is the essential human creativity that is human responsibility. The criticism of Blake's post-Apocalypse 'knowing' is that, unwittingly, illicitly and gratuitously, he (on behalf of mankind) abdicates human responsibility, and does that in offering to imagine its justifying triumph—the achieved goal, the τέλος, of the creative battle. There is no paradox in invoking Blake as I do, while declining to endorse his Jerusalem, or any suggestion that the creative human commitment to reality to which he pays his cogent testimony entails the kind of preoccupation with τέλος, or final cause, that belongs to the period side of him. Indeed, it is necessary in general, for the sake of the human heritage, to save Blake from the Blake authorities who make his genius and importance matters, essentially, of his relation to Boehme, Paracelsus, Plotinus and the 'perennial philosophy'—or some arcane tradition of esoteric truth that lives spent in research may recover for us.

This anti-industrial conviction is incidental to my great

reason for insisting on Blake. To associate his name with my theme and attitude is to emphasize that I *have* a positive theme, and that it and my attitude are truly positive —that (*pace* the accepted convention of dismissal) I am not a satiric polemist who takes a cruel and wanton pleasure in attacking 'poor Charles', Lord Annan, Lord Todd and the other well-deserving eminences. I have used the word 'creative' a great deal, but my problem has been to charge it with the crucial significance of the truth it portends—the truth about human nature and the human world. It must be obvious enough why I should have turned for help to Blake. There is that intense belief in life which entails the insistence on the individual, and there is the way in which Blake enforces the point that Lawrence, faced with egalitarian enlightenment and Bertrand Russell, was (as Birkin) to attempt to bring home to that benevolent philosophic mind with the word 'disquality'. And, at the opening of the line that runs (one can point out) to Lawrence through Dickens, there is the potent authority with which Blake conveys his knowledge that in creative work he himself serves something authoritative—a living reality that is not his selfhood: testimony that goes with his vividly imparted sense of the intrinsic relation between creativity in the artist and that which is inseparable from life.

But life in its continued livingness is what, of its very nature, can't be convincingly imagined in terms of a final cause vindicated in an achieved ultimate goal, for livingness is creativity, and creativity manifests itself in emerging newness. Eliot—though a poet—finds it impossible to believe in human creativity. But in Blake, the antithesis of Eliot, the belief in life is so strong that he can suppose that, if his art were equal to it, his Jerusalem might be evoked as the supreme manifestation of the livingness his own individual life's creativity has participated in, demonstrated and served. It belongs with our indebtedness to Blake that he prompts the distinguishing and limiting criticism entailed in a recognition of just what it is we owe him—owe to the vindicator of the creativity of life as, in man, responsibility.

I am myself helped to make the necessary points by Michael Polanyi, who, for years a Professor of Physical Chemistry, isn't thinking of Blake and won't be accused of 'literarism'. There can be no question of my offering to summarize him, and the sentences of his I quote have my context. But they seem to me congenial to it, and the reader who cares to do so can look up in the collection, *Knowing and Being*, the essays they come from. There they will find in Polanyi's own context a theory of knowledge (with, of course, ontological implications) that is closely and cogently argued in terms of evidence from a diversity of experimental fields. I adduce Polanyi in this way because support intrinsically so impressive from an intellectual background so different from mine makes the familiar offer to discredit my own argument by referring to something called 'literary' bias less plausible.

With his eye on the fallaciousness of positivism he writes ('Sense-giving and Sense-reading'[1]):

> The ideal of a strictly explicit knowledge is indeed self-contradictory; deprived of their tacit coefficients, all spoken words, all formulae, all maps and graphs, are strictly meaningless. An exact mathematical theory means nothing unless we recognize an inexact non-mathematical knowledge on which it bears and a person whose judgment upholds this bearing.

Polanyi's focal preoccupation is with the nature and

[1] *Knowing and Being*, page 195.

process of scientific discovery. The conclusion he presents is that the drive to discovery comes from a 'faculty for integrating signs of potentialities, a faculty that we may call the power of *anticipatory intuition*'. In discussing this power he finds it natural to use the word 'imagination':

> Poincaré emphasizes that illumination does not come without the previous work of the imagination. This applies also to what I call intuition. A problem for inquiry comes to the scientist in response to his roaming vision of yet undiscovered possibilities. Having chosen a problem, he thrusts his imagination forward in search of clues and the material he thus digs up—whether by speculation or experiment—is integrated by intuition into new surmises, and so the inquiry goes on to the end.'[1]

What Polanyi is investigating are the processes of what, with his attention on the field of scientific discovery, he calls 'unspecifiable knowledge', 'tacit knowledge' and 'tacit inference'. That there is a bearing on the field of my interest in Blake is plain; it is manifest in the way in which Polanyi eliminates the Cartesian dualism. The creative activity on the part of the mind, which could not so act if it were not incarnate in a living individual body, makes possible the illimitably advancing discovery of the physicist's reality. I say 'the physicist's' in order that I may go on to note that the argument is at the same time, and necessarily, a constatation of the complementary truth: the human mind can't be less really real than inanimate nature. It is of course an obvious truth, and few sane people doubt it, but in the world that produced the philosopher who assured me that a computer can write a poem we can't with complete conviction say that most people positively and unequivocally *believe* it. That is why I think it worth while for my present purpose to invoke Polanyi

[1] *Knowing and Being*, page 201.

as well as Blake. Polanyi's preoccupation with epistemology and ontology is a lively concern for human creative
activity and human responsibility (a word he uses and
emphasizes)—and 'lively' means 'in and of our time'. He
puts the truth I have just referred to with a pregnant
directness calculated to make an impact:

> If all men were exterminated, this would not affect the laws
> of nature. But the construction of machines would stop, and
> not until men arose again could machines be formed once
> more.[1]

It follows that neither the scientist's nor the technolo-
gist's concern with reality is self-sufficient—a consequence
to which Polanyi is fully alive. He doesn't, it is very plain,
need telling that the cultural collaboration that has created
the human world has had for the drive that sustains it a
not less essential concern for reality, and that that concern
is prior and can't be let die away without human disaster.

At this point I turn to Blake again, for Blake's preoccupation, that being his importance for us today, was
with man's non-Urizenic creativity—I think, when I use
that adjective, at one and the same time of Blake's sternly
godlike Newton and of the bearded wind-blown demiurge,
both with open compasses in the downstretched hand.
Not that Urizen hadn't his place, or part, in Blake's postulated total man, but as a figure in the myth he is the enemy,
or crucial aberration. Polanyi's importance is that, while
for him scientific discovery is the focal interest, he lends
himself (he is very accessible) so consciously and so effectively to the insistence that in the human world of the non-
Urizenic creativity, that which creates a language, the
concern is equally for the real: the criteria of reality are

---

[1] *Knowing and Being*, page 225. ('Life's Irreducible Structure.')

the same.[1] The considerations that dispose of the Cartesian dualism (and its sequelae in our time) entail that constatation. And the argument moves from the sciences of inanimate nature in an unbroken transition to this:

> Thus our understanding of living beings involves at all levels a measure of indwelling; our interest in life is always convivial. There is no break therefore in passing from biology to the acceptance of our cultural calling in which we share the life of a human society, including the life of its ancestors, the authors of our cultural heritage.[2]

To recognize the importance of the cultural heritage in that way goes with the strength of Polanyi's thought. We are left, however, with the phrase, 'cultural heritage', and to get beyond it, and what it suggests, to the conception of a continuously living culture the continuity of which is collaborative creativity we must have recourse to our own fresh observation and insight. And thought about Blake, the great poetic innovator in a long traditional line, ought to generate at least a readiness for the crucial recognition. But the cue for the word 'collaborative' doesn't leap at us out of Blake's work. How could it? His mythopoeia, in its complexities of profusion and change, depends for its coherence and its intelligibility (such as they are for the

[1] There would be no point in my offering to summarize Polanyi's definition of them. The defining is done, both explicitly and implicitly, in many contexts that bring light and strengthen cogency. I will quote only this:

> 'Our capacity to endow language with meaning must be recognized as a particular instance of our sense-giving powers. We must realize that to use language is a performance of the same kind as our integration of visual clues for perceiving an object, or as the viewing of a stereo picture, or our integration of muscular contractions in walking or driving a motor car, or as the conducting of a game of chess . . .'
> *Knowing and Being*, page 193.

[2] *Knowing and Being*, page 136.

reader) on the commanding presence in our minds of the
Eternal Man. If we forget this (as we are bound to do—
and we can never fully realize the presence) we are lost;
for the drama of personified aspects or faculties is confusing, and the unschematic personal manifestations of
the characters, though welcome enough, tend to be rout-
ing. Even when what the reversal of the Fall gives us
appears as Jerusalem (a city as well as a person), we are
no nearer to recovering those particular identities, those
interacting human individuals, in whom reality resided for
Blake himself—Blake the poet and artist.

The ambition to possess an achieved knowledge of ultimate solutions and ultimate goals is neither for poets and
artists nor for those who tackle the human problem at the
level I here propose for my own attempts. Polanyi, his
main concern being with scientific discovery, discusses
this as a specialized mode of the process by which man
habitually comes to terms with the real world in which he
lives. The scientist's anticipatory apprehension of aspects
of reality that are still unknown to him and unspecifiable
is a developed and cultivated form of a characteristic that
is distinctive of life in general. I, in discussing creative
writers and that movement towards the achieved work (to
say 'conception' would be misleading) which starts with
an elusive sense of some coherence or pattern to be found
in experience, or the sense of some deep inner need, have
found myself using the words '*ahnung*' and 'nisus'.
Recently, as student collaborators will testify, I have re-
sorted to them a great deal in the discussion of Eliot's
religious poetry: the movement from the avowal of utter
destitution in 'The Hollow Men' and the recognition of
the nature of the need—need so exacting in its intensity
that it imposes the astonishingly rendered continence of

affirmation—through *Four Quartets* to the word 'Incarnation' in 'The Dry Salvages'. It mustn't be taken as implying that what Eliot offers as an apprehension of the transcendent spiritual reality satisfies. It seems to me that, for all his slowness to affirm and his subtly elaborate technique of exploration and definition, he plays false with the *ahnung* and fails to protect the nisus against the promptings of the selfhood. The failure is a failure of the courage of self-knowledge—that is, a weakness of the disinterested 'identity', and the consequence manifests itself in the poem as a basic nullifying contradiction.

There is, I have suggested, pointing to the symbolic Jerusalem, contradiction in Blake, but self-ignorance will hardly be predicated of him. His insight into the human psyche began where it was most intimately open to him, and he had the courage to recognize what he saw (a point that Eliot makes in his curious and self-revealing essay on Blake).[1] His searching insight into the complexities and subtleties of essential humanity is what makes him so important to us, and I have adduced him to enforce my emphasis on the creativity that is human responsibility, and the maladies that ensue when creativity is frustrated and implicitly denied. The contradiction presents itself in the offered *terminus ad quem*; for the τέλος, Jerusalem, *is* an end: to the evoking of that unimaginable consummation Blake's creativity is ill-adapted; the artist, whatever he may suppose, can't believe in it—can't in the artist's way.

In making this criticism one is paying an implicit tribute to Blake, for the thought that brings one to it represents an essential indebtedness to him. I express my sense of

---

[1] 'And because he was not distracted or frightened, or occupied in anything but exact statements, he understood. He was naked, and saw man naked, and from the centre of his own crystal.'

this when I choose my Blakean title, though not committed—and telling myself so—to anything in the nature of building Jerusalem. My focal preoccupation in the following pages is with the creating of the university—a very different matter. The university as I contend for it is not an ultimate human goal; it is the answer to a present extremely urgent need of civilization. The need is to find a way to save cultural continuity, that continuous collaborative renewal which keeps the 'heritage' of perception, judgment, responsibility and spiritual awareness alive, responsive to change, and authoritative for guidance. In that continuity inhere the *ahnung* and the nisus that, effectively energizing the few, represent humanity's chance of escaping the disasters from which scientists, technologists and economists, as such and alone, cannot save us. For not only faith, but creativity in the realm of the spirit, the realm of human significances, is necessary.

The pair of unvernacular words I have used should get what further definition they may seem to need from the following discourses, which were conceived as forming, in their totality, a relevant context. I will only say here that Blake, when he offers his Jerusalem as the goal achieved, assumes too easily that his creative nisus has done the work that was its significance and may rest from now on. It is in and of the situation not only of a Blake but of humanity (in the personal foci that represent it) that, the problem of the τέλος, the ultimate goal, being in question, the nisus can never achieve the final satisfaction and the supersession of the *ahnung*; what it *can* hope to achieve is the advance that brings a fresh field of data into view and establishes a new situation for insight to work on.

I am not for a moment, I had better add, presuming to relegate philosophical, cosmological and theological thought to a category of the unnecessary. Very consciously at this present moment of history, and addressing myself to a problem that for all persons capable of real imaginative responsibility is an urgent challenge, I am defining a given kind of effort that, as the necessary answer, has to be made, and I have distributed my emphasis accordingly. The university that represents this effort requires, as I hope to have made plain in these discourses, the co-presence of the major disciplines of knowledge, inquiry and understanding, and these are likely to have their parts in any cultural nisus generated. And, where religion is in question, that I stop with the adducing of the Laurentian formulation dramatized in Tom Brangwen will surely not be misunderstood.

My argument was necessarily a complex one; I had to relate—which sometimes meant merely touching on them —a diversity of considerations in such a way as to bring out the significant cogency of my contentions and the urgency of the crisis they concerned. I come here to the form in which my argument is presented—I mean the series of lectures constituting this book. It is true that the opening discourse was (in the sense exemplified in Johnson's Life of Dryden) occasional. I was invited by the Amalgamation Club at Downing to give the annual Richmond lecture, and, having for some while had a steadily reinforced charge of reason for thinking that the famous Rede lecture represented a theme that badly needed public attention, I took the opportunity. The storm of indignation, protracted and abusive (of me), that followed the appearance of my lecture in the *Spectator* (which had asked for it) brought home to me that very much fuller treatment of the whole complexity of issues involved was urgently called for. I gave 'Luddites? or There is only one culture' in its first improvised form at the invitation of the students of Churchill College, with which Lord Snow is closely associated. By the time I had accepted the invitation to address at Gregynog the staffs of the English Departments of the University of Wales, I realized that I had hit on the way of writing the book that seemed to me so badly needed—the book answering to my very much 'engaged' sense of the situation. The problem was to combine a clear and compelling recognition of the range, diversity and depth of the issues—the complexity that attended the momentousness—with the necessary disturbing directness of attack. I say 'disturbing directness' because of the need to precipitate the kind of active conviction which that phrase suggests. Each lecture was conceived accordingly as having a certain completeness in itself—and that it told in that way was strongly suggested by the very gratifying response. The audience, on the other hand, couldn't fail to recognize a large possibility of development and the clear implication of a supporting context that asked to be, and could be, produced. The series, however, was not heard as a series by any audience. But the author himself was in a position to appreciate the way in which the conception of a comprehensive whole to be achieved and, at the completion of the series, apparent, facilitated the business of making each of the constituent lectures itself in its way an effective whole.

The book as a book needed that kind of self-sufficiency in the parts if it was to serve the intended function—that is, to have the urgent immediacy that one doesn't associate with the methodically marshalled argument of a treatise. The lecture circumstances helped me immensely to deal with the paralysing problem that faces everyone who is committed to presenting a complex theme—and any important argument entails that problem—in written exposition: one paragraph comes after another, one chapter after another; one is committed to sequence and succession —to presenting the diversity of considerations separately and at a distance from the focal passages in which the significance they together portend gets its weighed and calculated statement. Each lecture was a new plunge and a fresh particular approach, prompted by the chances of *actualité*: the emphasis was different, and repeated thematic material had a different context and belonged to a different organization. Repetition in itself served an essential function, impressing, in a diversity of contexts, key-considerations on the memory. The repeated adducing of certain decisive names—for instance, Blake, Dickens, Lawrence and Shakespeare—had, for a public exposed to the higher name-dropping and the practice of poly-intellectual smatter, its point.

For the emphasis represented by the invoking of Blake's name in the title of this book I have indicated some of the reasons. Generally, there is the need (absurd as I find it) to insist that my contention can't be honestly reduced to destructive polemic; that, on the contrary, my concern is as profoundly positive as any concern could be, and is given in the response I make to the statesman's maxim, 'Politics is the art of the possible'—a response that is more than a statement: 'We create possibility'. More particularly, I think of a description of myself I read first in *Corriere della Sera,* and now come on again in a book:[1] 'puritano frenetico'. The Italian journalist (whom I have never met) obviously picked up the account of me he registers in that phrase in the London literary or academico-journalistic world. There is an irony for me in recalling that in the 1930s the case for the steady aca-

[1] *Sessanta Posizioni*, Alberto Arbasino (Feltrinelli).

demic disfavour I enjoyed was 'We don't like the books he lends undergraduates'. What that insinuated was bluntly expressed as a safe quip in *The Granta* (where a manly Faculty potentate—he was also a towpath coach—was, it was well-known, a congenial and prompting intimate): 'The Leavis Prize for Pornography'.

The accepted attitude towards Lawrence, of course, changed. When permissiveness, having become general, rallied with so massive a show of moral conviction to the support of Sir Allen Lane's knight-errantry and achieved the historic triumph, the literary world recalled my pioneering record as Lawrence's critical advocate—recalled it (that was the idea) to my confusion. Had I not backed Lawrence then? For my refusal to lend myself to the campaign that made *Lady Chatterley's Lover* a fabulous bestseller, or to express anything but dismay and apprehension at the consequences, discreditably petty motives were imputed. But in the new climate of victorious permissiveness the truly discreditable characterization to have earned is that which, my name having come up, was impressed on Signor Arbasino: 'puritan'. Permissiveness, however, doesn't confine itself to the province of sex, and there is a kind of challenge, or offensiveness, that incurs a much stronger resentment than that which meets the sour perversities of questioning aimed at our new and rapidly achieved sexual emancipation. What indeed is registered in the account picked up by Signor Arbasino in his London frequentations is the hatred generated among the enlightened by the insistent reminder that there are criteria to be observed not less—certainly not less—in the centrally and basically human non-specialist field of intelligence, concern and judgment than elsewhere, and that, with an authoritatively critical bearing on all utterance that asks to be taken seriously, there is something of the first importance that the modish literary world neither loves nor is familiar with—something that here I can only gesture towards with the triad: 'disinterestedness, essential human responsibility, and grounded conviction'. To enforce my intimation that all three members are necessary to the identifying of what I intend, I can say: 'Think of Daniel Doyce.'[1] In him Dickens gives us the antithesis of what Lord Annan calls 'pluralism', the term he chooses for his own attitude or habit. 'Pluralism' denotes a sitting-easy to questions of responsibility, intellectual standard, and even superficial consistency, the aplomb, or suppleness, being conditioned by a coterie-confident sense of one's own unquestioned sufficiency—or superiority. To become known as given, impertinently, importunately and unanswerably, to exposing that unexacting ethos *d'élite* for what it is, and, challenged, is conscious of being, is to incur the character that Signor Arbasino relays to Italy: *puritano frenetico*.

No one would in any context call Blake a puritan. He warred angrily against codes and conventions that favoured repressive righteousness or the association of sex with shame. On the other hand, it is impossible to doubt that to what the editor of the *New Statesman* advocates as 'the democratization of sexual pleasure'[2] he would have reacted

[1] See *Dickens the Novelist*, pages 238-9.

[2] 1 May 1971: 'The hypocrisy, moreover, is not confined to money. Lord Longford and his supporters have turned it into a class issue. For years they were prepared to tolerate a society where pornography was available to the upper-middle class and the aristocracy, but kept discreetly hidden from the public gaze. What they cannot tolerate, it seems, is the democratization of sexual pleasure.'

The irresponsibility, reductive *élan* and uninhibited logic of this (the Editor wouldn't be embarrassed by one's comment on the implicit assimilation of 'sexual pleasure' to pornography) are not countered or mitigated by anything else in the editorial. Far from it.

with fierce, contemptuous and uncompromising enmity. His sense of the vital significance of sex—and no one has ever had a fuller apprehension of what 'vital' implies, his insight (that is) into sex as something more than a specific marked-off field of potential physical pleasure or pornographically imaginative addiction, went with, or was, his insight into the complexities and delicacies of love, and the crucial necessity of love to life.

But there is a dawning unselfrecognized conviction that we can get on, and get on better, without much life; and that is the most frightening thing about our civilization. The non-recognition derives from the insidious way in which, depending as we do on mysterious mechanisms—for most of us they mainly *are* that—among which we live, and which make our external civilization possible, we are inevitably unable to separate the human use of the mechanisms from the being used by them.

Of course, they can't literally use us; they depend on human initiative and intelligence, a truism put in his characteristic way by Polanyi ('If all men were exterminated . . . the production of machines would stop'). One can't help thinking that no one would with sober conviction question this, but there is depressing evidence that in the world we live in it is not what I mean by conviction in the important sense that determines decisive assumption and attitude, but something very different—something that, if we look for the signs of real and conscious human responsibility, we have to call inertness. True, I have twice received communications from a computer enthusiast bringing to my notice the unquestionable fact that *homo sapiens* is obsolete, and intimating (I gathered) that it was my clear duty to cooperate in speeding up his supersession. The first time, my reaction was 'O, a lunatic';

but on reflection I realized that I was faced with a forceful expression of an implicit kind of conviction (naïve, but not abnormal, it may be called—it is at any rate unquestioning) that does in fact characterize our civilization. It gives us the cultural climate that makes it possible for educated persons to assure one that a computer can write a poem. Such a climate obviously doesn't foster, or favour, aliveness to the real menace represented by my acclaimer of the cybernetic future. It is not that the super-session of *homo sapiens* will be generally and joyfully recognized as inevitable. It is rather that the kind of resolute statistico-egalitarian reductivism so zealously advocated by Mr Christopher Price in education will triumph, the progress towards that goal being, by the enlightened, acclaimed as all in keeping with man's accelerating conquest of nature.

It would be pointless of course to think of adducing Blake for the persuasion of politicians or the mass of the enlightened. But my concern is with the university, and the focal audience, if there *is* one, will be in the universities as they are, and especially in English departments. There professional self-sufficiency, mere habit and coterie-conceit might seem to dominate. It would at all events be the extreme of unrealism to hope for an ardent general response to the most persuasive appeal. But, as I have said, creative change is not initiated by majorities, and that truth holds of the personnel within the departmental frontiers. There will be those—sustained, perhaps, by communication across the frontiers—who don't need conversion; there will be more whom the sharpening of perception may awaken to conscious and convinced responsibility; there is the sense of collaborative purpose to be aroused and strengthened. And though the ambition- and success-hardened are conscienceless and, in their will to power, prestige and privilege, formidably coagulative, the challenge to them to respect their calling and honour the beliefs that, in living by it, they implicitly profess isn't necessarily futile: there may be, if not shame, embarrassment and a measure of inhibition. Further—or it should, perhaps, be firstly—there is the need to sustain the courage of one's own conviction; the thought of Blake as a source of reassurance and resolution gets its prompting there.

In this avowal I confess that the prospect is discouraging. It is necessary to look this fact in the face, and to insist on it—in order to go on to say: 'The resolution must be according'. I have in the following lectures dwelt on the way in which the civilization we live in and which has conquered the world makes it almost impossible to get recognition for the very nature of the human needs it ignores and thwarts. The fact that the needs are basic and that their frustration manifests itself more and more in ugly and frightening forms of dissatisfaction on a scale that stultifies the optimism of the enlightened forbids us, I argue, to regard the creative effort as futile—the effort, in the first place, to get them duly recognized for what they are. In any case, the very belief in life precludes an abandonment of the battle, or the thought that it must be lost.

But the prospect before us grows more menacing. The great theme in the newspapers can only intensify one's sense of the urgency. The arguments for 'becoming European' may be strong enough to carry the day—may even deserve to do so, but they take no cognizance of the issues with which this book is concerned and they implicitly threaten the human cause that insists on such issues. It is plain enough that when (or if) this country becomes part of an integrated Europe the mechanisms, magnitudes and blanknesses in the way of getting the essential human problem attended to will become more formidable. It is not for me to pronounce between the for and the against of the economico-political debate. Whatever the outcome, the need for conviction and resolution and indefatigable battle in regard to issues that, if noticed at all, will be in general dismissed as without moment or substantial meaning will remain and will not lessen. On the prospect, then, of our 'entering Europe' the arguments involved in the preoccupation of this book have certainly a bearing, and the perception of the relevance can't but manifest itself in an even stronger sense that there is no time to lose. If the very idea of the necessary effort is not to die away in the vast impercipience, the public, or community, in which alone perception and purpose can live must become conscious of itself as such now, and of its responsibilities, and see to the forming, reinforcing and multiplying of its vital—its generative—centres.

I know, of course, that the old kinds of misrepresentation will go on; that I shall be represented as contending that literary cultivation will save us. And by the politician's criteria the suggestion that the university can matter in the way I argue is absurd. But it is not absurd, though it may seem desperate. In any case, the conceptions involved are new and unfamiliar, being a creative response to change—as cultural continuity itself is. That there *has* been swift and radical change, and that more is upon us, no one will question, and there is a general recognition, helpless enough, of a deep and frightening human disorder (the consequence, or concomitant, of change) that menaces civilization itself. Further, the recognition goes with an uneasy sense that this disorder is of a kind that economic reforms, a higher 'standard of living', and increased permissiveness—even accompanied by the elimination of venereal disease and every inhibition, will not cure. In such a world perception and conviction have their responsibility, and it has yet to be proved that there is not a potential minority-public to be rallied, a community of the convinced to be made conscious of itself, large enough to tell disproportionately in relation to its statistical presence.

Of the following lectures, the second ('Luddites') was given in 1966 at the Universities of Cornell and Harvard; the third ('English, Unrest and Continuity') in 1969 at Gregynog, to an audience from the University of Wales; the fourth, in 1970, during my tenure of the Churchill Visiting Professorship at the University of Bristol; and the closing two in 1970 and 1971 at the University of York.

I have to acknowledge my indebtedness to the *Times Literary Supplement* for permission to reprint those given in Wales and at Bristol, and to *The Human World* for permission to reprint those given at York. It is only fair to add that the *Times Literary Supplement* printed rejoinders by Lord Annan and Lord Snow in the issues of April 30th and July 9th 1970 respectively.

II
Two Cultures?
The Significance of Lord Snow

TWO CULTURES?
THE SIGNIFICANCE OF
LORD SNOW

IF confidence in oneself as a master-mind, qualified by capacity, insight and knowledge to pronounce authoritatively on the frightening problems of our civilization, is genius, then there can be no doubt about Sir Charles Snow's. He has no hesitations. Of course, anyone who offers to speak with inwardness and authority on both science and literature will be conscious of more than ordinary powers, but one can imagine such consciousness going with a certain modesty—with a strong sense, indeed, of a limited range and a limited warrant. The peculiar quality of Snow's assurance expresses itself in a pervasive tone; a tone of which one can say that, while only genius could justify it, one cannot readily think of genius adopting it. It is the tone we have (in so far as it can be given in an isolated sentence) here:

> The only writer of world-class who seems to have had an understanding of the industrial revolution was Ibsen in his old age: and there wasn't much that old man didn't understand.

Clearly, there is still less Sir Charles Snow doesn't understand: he pays the tribute with authority. We take the implication and take it more surely at its full value because it carries the *élan*, the essential inspiration, of the whole self-assured performance. Yet Snow is in fact portentously ignorant. No doubt he could himself pass with ease the tests he proposes for his literary friends with the intimation that *they* would fail them, and so expose themselves as deplorably less well educated in respect of science

41

than he, though a scientist, can claim to be in respect of literature. I have no doubt that *he* can define a machine-tool and state the second law of thermodynamics. It is even possible, I suppose (though I am obliged to say that the evidence seems to me to be against it), that he could make a plausible show of being inward with the Contra-diction of Parity, that esoteric upshot of highly subtle experiment which, he suggests, if things were well with our education, would have been a major topic at our High Tables. But of history, of the nature of civilization and of the history of its recent developments, of the human history of the Industrial Revolution, of the human signifi-cances entailed in that revolution, of literature, of the nature of that kind of collaborative human creativity of which literature is the type, it is hardly an exaggeration to say that Snow exposes complacently a complete ig-norance.

The judgment I have to come out with is that not only is he not a genius; he is intellectually as undistinguished as it is possible to be. If that were all, and Snow were merely negligible, there would be no need to say so in any insistent public way, and one wouldn't choose to do it. But I used the adverb 'portentously' just now with full intention: Snow is a portent. He is a portent in that, being in himself negligible, he has become for a vast public on both sides of the Atlantic a master-mind and a sage. His significance is that he has been accepted—or perhaps the point is better made by saying 'created': he has been created as authoritative intellect by the cultural conditions manifested in his acceptance. Really distinguished minds are themselves, of course, *of* their age; they are responsive at the deepest level to its peculiar strains and challenges: that is why they are able to be truly illuminating and

prophetic and to influence the world positively and creatively. Snow's relation to the age is of a different kind;
it is characterized not by insight and spiritual energy, but
by blindness, unconsciousness and automatism. He doesn't
know what he means, and doesn't know he doesn't know.
That is what his intoxicating sense of a message and a
public function, his inspiration, amounts to. It is not any
challenge he thinks of himself as uttering, but the challenge
he *is*, that demands our attention. The commentary I have
to make on him is necessarily drastic and dismissive; but
don't, I beg, suppose that I am enjoying a slaughterous
field-day. Snow, I repeat, is in himself negligible. My preoccupation is positive in spirit. Snow points to its nature
when he turns his wisdom upon education and the university.

I have not been quick to propose for myself the duty
of dealing with him: that will, I hope, be granted. *The
Two Cultures and the Scientific Revolution*, the Rede lecture
which established him as an Intellect and a Sage, was given
at this ancient university in 1959. I turned over the pages
of the printed lecture in the show-room of the Cambridge
University Press, was struck by the mode of expression
Snow found proper and natural, perceived plainly enough
what kind of performance the lecture was, and had no
inclination to lay down three and sixpence. To my sur-
prise, however, it rapidly took on the standing of a classic.
It was continually being referred to—and not only in the
Sunday papers—as if Snow, that rarely qualified and profoundly original mind, had given trenchant formulation
to a key contemporary truth. What brought me to see
that I must overcome the inner protest, and pay my three
and sixpence, was the realizing, from marking scholarship
scripts, that sixth-form masters were making their bright

boys read Snow as doctrinal, definitive and formative—
and a good examination investment.

Well, I bought the lecture last summer, and, having
noted that it had reached the sixth printing, read it through.
I was then for the first time in a position to know how
mild a statement it is to say that *The Two Cultures* exhibits
an utter lack of intellectual distinction and an embarrassing
vulgarity of style. The lecture, in fact, with its show of
giving us the easily controlled spontaneity of the great
man's talk, exemplifies kinds of bad writing in such rich-
ness and so significant a way that there would, I grant,
be some point in the schoolmaster's using it as a text for
elementary criticism; criticism of the style, here, becomes,
as it follows down into analysis, criticism of the thought,
the essence, the pretensions.

The intellectual nullity is what constitutes any difficulty
there may be in dealing with Snow's panoptic pseudo-
cogencies, his parade of a thesis: a mind to be argued
with—that is not there; what we have is something other.
Take that crucial term 'culture', without which and the
work he relies on it to do for him Snow would be deprived
of his seer's profundity and his show of a message. His
use of it focuses for us (if I may be permitted what seems
to me an apt paradox) the intellectual nullity; it confronts
us unmistakably with the absence of the thought that is
capable of posing problems (let alone answering them). The
general nature of his position and his claim to authority
are well known: there are the two uncommunicating and
mutually indifferent cultures, there is the need to bring
them together, and there is C. P. Snow, whose place in
history is that he has them both, so that we have in him
the paradigm of the desired and necessary union.

Snow is, of course, a—no, I can't say that; he isn't;

Snow thinks of himself as a novelist. I don't want to discuss that aspect of him, but I can't avoid saying something. The widespread belief that he is a distinguished novelist (and that it should be widespread is significant of the conditions that produced him) has certainly its part in the success with which he has got himself accepted as a mind. The seriousness with which he takes himself as a novelist is complete—if seriousness can be so ineffably blank, so unaware. Explaining why he should have cut short a brilliant career (we are to understand) as a scientist, he tells us that it had always been his vocation to be a writer. And he assumes with a happy and undoubting matter-of-factness—the signs are unmistakable—that his sense of vocation has been triumphantly vindicated and that he is beyond question a novelist of a high order (of 'world-class' even, to adopt his own idiom). Confidence so astonishingly enjoyed might politely be called memorable —if one could imagine the memory of Snow the novelist long persisting; but it won't, it can't, in spite of the British Council's brochure on him (he is a British Council classic). I say 'astonishingly enjoyed', for as a novelist he doesn't exist; he doesn't begin to exist. He can't be said to know what a novel is. The nonenity is apparent on every page of his fictions—consistently manifested, whatever aspect of a novel one looks for. I am trying to remember where I heard (can I have dreamed it?) that they are composed for him by an electronic brain called Charlie, into which the instructions are fed in the form of the chapter-headings. However that may be, he—or the brain (if that's the explanation)—can't do any of the things the power to do which makes a novelist. He tells you what you are to take him as doing, but he can give you no more than the telling. When the characters are supposed to fall in love you are told they do, but he can't show it happening. Abundant dialogue assures you that this is the novelistic art, but never was dialogue more inept; to imagine it spoken is impossible. And Snow is helpless to suggest character in speech. He announces in his chapter-headings the themes and developments in which we are to see the significance of what follows, but what follows adds nothing to the effect of the announcement, and there is no more significance in the completed book than there is drama—or life. It is not merely that Snow can't make his characters live for us—that he lacks *that* creative power; the characters as he thinks of them are so impoverished in the interests they are supposed to have and to represent that even if they had been made to live, one would have asked of them, individually and in the lump: 'What of life is there here, and what significance capable of engaging an educated mind *could* be conveyed through such representatives of humanity?'

Among the most current novels of Snow's are those which offer to depict from the inside the senior academic world of Cambridge, and they suggest as characteristic of that world lives and dominant interests of such unrelieved and cultureless banality that, if one could credit Snow's art with any power of imaginative impact, one would say that he had done his university much harm— for this is a time when the image of the ancient university that is entertained at large matters immensely. Even when he makes a suspect piece of research central to his plot, as in that feeble exercise, *The Affair*, he does no more than a very incompetent manufacturer of whodunnits could do: no corresponding intellectual interest comes into the novel; science is a mere word, the vocation merely postulated. It didn't take a brilliant research scientist to deal with the alleged piece of research as Snow deals with it—or a scientist of any kind. Both George Eliot and Lawrence could have made such a theme incomparably more real.

What the novelist really believes in, the experience he identifies his profoundest ego with because it makes him feel himself a distinguished man and a lord of life, is given us in Lewis Eliot. Eliot has inhabited the Corridors of Power; that is what really matters; that is what qualifies him to look down upon these dons, the scientists as well as the literary intellectuals, with a genially 'placing' wisdom from above; there we have the actual Snow, who, I repeat, is a portent of our civilization; there we have the explanation of his confident sense of importance, which, in an extraordinary way, becomes where his writing is concerned a conviction of genius: he has known from inside the Corridors of Power. That he has really *been* a scientist, that science as such has ever, in any important inward way, existed for him, there is no evidence in his fiction.

And I have to say now that in *The Two Cultures and the Scientific Revolution* there is no evidence, either. The only presence science has is as a matter of external reference, entailed in a show of knowledgeableness. Of qualities that one might set to the credit of a scientific training there are none. As far as the internal evidence goes, the lecture was conceived and written by someone who had not had the advantage of an intellectual discipline of any kind. I was on the point of illustrating this truth from Snow's way with the term 'culture'—a term so important for his purposes. By way of enforcing his testimony that the scientists 'have their own culture', he tells us: 'This culture contains a great deal of argument, usually much more rigorous, and almost always at a higher conceptual level, than literary persons' arguments.' But the argument of Snow's Rede lecture is at an immensely *lower* conceptual level, and incomparably more loose and inconsequent, than any I myself, a literary person, should permit in a group discussion I was conducting, let alone a pupil's essay.

Thought, it is true, in the field in which Snow challenges us, doesn't admit of control by strict definition of the key terms; but the more fully one realizes this the more aware will one be of the need to cultivate a vigilant responsibility in using them, and an alert consciousness of any changes of force they may incur as the argument passes from context to context. And what I have to say is that Snow's argument proceeds with so extreme a *naïveté* of unconsciousness and irresponsibility that to call it a movement of thought is to flatter it.

Take the confident ease of his way with what he calls 'the Literary Culture', that one of his opposed pair which, as a novelist, he feels himself qualified to present to us with a peculiar personal authority. He identifies 'the Literary Culture' with, to use his own phrase, the 'literary intellectual'—by which he means the modish literary world; his 'intellectual' is the intellectual of the *New Statesman* circle and the reviewing in the Sunday papers. Snow accepts this 'culture' implicitly as the *haute culture* of our time; he takes it as representing the age's finer consciousness so far as a culture ignorant of science can. He, we are to understand, has it, and at the same time the scientific culture; he unites the two. I can't help remarking that this suggested equivalence (equivalence at any rate in reality) must constitute for me, a literary person, the gravest suspicion regarding the scientific one of Snow's two cultures. For his 'literary culture' is something that those genuinely interested in literature can only regard with contempt and resolute hostility. Snow's 'literary intellectual' is the enemy of art and life.

Note with what sublime, comic and frightening ease (for this sage is after all a Cambridge man) Snow, without any sense of there having been a shift, slips from his 'literary culture' into 'the traditional culture'. The feat of innocent unawareness is striking and significant enough when he is talking of the contemporary scene. But when, with the same ease, he carries the matter-of-fact identification into the past—'the traditional culture', he tells us, with reference to the Industrial Revolution, 'didn't notice: or when it did notice, didn't like what it saw'—the significance becomes so portentous as to be hardly credible. But Snow, we must remind ourselves, *is* frightening in his capacity of representative phenomenon. He knows nothing of history. He pronounces about it with as complete a confidence as he pronounces about literature (French, Russian and American as well as English), but he is equally ignorant of both. He has no notion of the changes in civilization that have produced his 'literary culture' and made it possible for C. P. Snow to enjoy a status of distinguished intellectual, have the encouragement of knowing that his Rede lecture is earnestly studied in sixth forms, and be (with practical consequences) an authority in the field of higher education: things that the real, the living, 'traditional culture' (for there is a reality answering to that phrase) can no more countenance today than it could have foreseen them in the nineteenth century.

The intellectual nullity apparent in his way with the term 'culture' is only emphasized for us when, coming to his other culture, that of the scientist, he makes, as himself a scientist, his odd show of a concern for a 'high conceptual level'. 'At one pole,' he says, 'the scientific culture really is a culture, not only in an intellectual, but also in an anthropological sense.' The offered justification for that 'anthropological sense' is given, we find, if we examine the context, in this sentence: 'Without thinking about it they respond alike.' Snow adds: 'That is what a culture means.' We needn't bother one way or the other about the 'anthropological'; what is certain is that Snow gives us here a hint worth taking up. He, of course, is supposed to be thinking, and thinking profoundly, in that Rede lecture, but actually it is a perfect document of the kind of 'culture', to use his word, that he here defines—defines, even though unconscious of the full significance of what he says, the formal definition getting its completion and charge from the whole context—that is, from the actual performance. His unconsciousness is an essential characteristic. 'Without thinking, they respond alike': Snow's habits as an intellectual and a sage were formed in such a milieu. Thinking is a difficult art and requires training and practice in any given field. It is a pathetic and comic —and menacing—illusion on Snow's part that he is capable of thought on the problems he offers to advise us on. If his lecture has any value for use in schools—or universities—it is as a document for the study of cliché.

We think of cliché commonly as a matter of style. But style is a habit of expression, and a habit of expression that runs to the cliché tells us something adverse about the quality of the thought expressed. 'History is merciless to failure': Snow makes play with a good many propositions of that kind—if 'proposition' is the right word. We call them clichés because, though Snow clearly feels that he is expressing thought, the thought, considered even for a moment, is seen to be a mere phantom, and Snow's illusion is due to the fact that he is *not* thinking, but resting inertly (though with a sense of power) on vague memories of the way in which he had heard (or seen) such phrases used. They carry for him—he belonging to what he calls a 'culture'—a charge of currency-value which is independent of first-hand, that is, actual, thinking. He would be surprised if he were told they are clichés.

He would be still more surprised to be told it is cliché when, describing the distinctive traits of his scientists, he says: 'they have the future in their bones'. He clearly feels that it has an idiosyncratic speech-raciness that gives his wisdom a genial authority. But it is basic cliché—for Snow's pretensions, more damagingly cliché than the kind of thing I instanced first, for it dismisses the issue, tacitly eliminates the problem, discussion of which would have been the *raison d'être* of the lecture if Snow had been capable of the preoccupation, and the accordant exercise of thought, he advertises.

Such a phrase as 'they have the future in their bones' (and Snow repeats it) cannot be explained as a meaningful proposition, and in that sense has no meaning. It emerges spontaneously from the cultural world to which Snow belongs and it registers uncritically (hence the self-evident force it has for him) its assumptions and attitudes and ignorances. That world, I was on the point of saying, is the world of his 'scientific culture', but I might equally have said that it is the world of the *New Statesman*, the *Guardian* and the Sunday papers. And Snow rides on an advancing swell of cliché: this exhilarating motion is what he takes for inspired and authoritative thought.

He brings out the intended commendatory force, and the actual large significance, of 'they have the future in their bones' (there is nothing else by way of clarification) by telling us antithetically of the representatives of 'the

traditional culture': 'they are natural Luddites'. It is a
general charge, and he makes quite plain that he includes
in it the creators of English literature in the nineteenth
century and the twentieth. The upshot is that if you insist
on the need for any other kind of concern, entailing forethought, action and provision, about the human future—
any other kind of misgiving—than that which talks in
terms of productivity, material standards of living, hygienic
and technological progress, then you are a Luddite. Snow's
position, for all the mess of clichés and sentimental banali-
ties that constitutes his style, is unequivocal.

It might seem an odd position for one who proudly
thinks of himself as a major novelist. But I now come to
the point when I have again to say, with a more sharply
focused intention this time, that Snow not only hasn't in
him the beginnings of a novelist; he is utterly without a
glimmer of what creative literature is, or why it matters.
That significant truth comes home to us, amusingly but
finally, when, near his opening, he makes a point of impressing on us that, as himself a creative writer, he is
humanly (shall I say)? supremely well qualified—that he
emphatically *has* a soul. 'The individual condition of each
of us,' he tells us, 'is tragic,' and, by way of explaining
that statement, he adds, 'we die alone'. Once he says 'we
live alone', but in general—for he makes his point redundantly—he prefers to stress dying; it's more solemn. He
is enforcing a superiority to be recognized in the scientists: they, he says, 'see no reason why, just because the
individual condition is tragic, so must the social condition
be'. For himself, with tragic stoicism, he says, 'we die
alone: all right,' but—which is his message, the sum of
his wisdom—'there is social hope'.

He is repetitious, but he develops no explanation further than this. It doesn't occur to him that there is any need, stultifying as anyone capable of thought can see the antithesis to be. What *is* the 'social condition' that has nothing to do with the 'individual condition'? What is the 'social hope' that transcends, cancels or makes indifferent the inescapable tragic condition of each individual? Where, if not in individuals, is what is hoped for—a *non*-tragic condition, one supposes—to be located? Or are we to find the reality of life in hoping for other people a kind of felicity about which as proposed for ourselves ('jam', Snow calls it later—we die alone, but there's jam to be had first) we have no illusions? Snow's pompous phrases give us the central and supreme instance of what I have called 'basic cliché'. He takes over inertly—takes over as a self-evident simple clarity—the characteristic and disastrous confusion of the civilization he is offering to instruct.

It is a confusion to which all creative writers are tacit enemies. The greatest English writer of our century dealt with it explicitly—dealt with it again and again, in many ways, and left to our hand what should be the classical exposure. But Snow, exhibiting his inwardness with modern literature by enumerating the writers who above all matter, leaves Lawrence out (though he offers us Wyndham Lewis —the brutal and boring Wyndham Lewis). Lawrence, intent with all his being on the nature and movement of the civilization of the West, turned the intelligence of genius on what I have called the characteristic confusion. He diagnoses it in his supreme novel, *Women in Love*, both discursively and by the poetic means of a great novelist. Concerned with enforcing in relation to what may be called a quintessential presentment of the modern world the Laurentian maxim that 'nothing matters but life', he insists on the truth that only in living individuals is life

53

there, and individual lives cannot be aggregated or equated or dealt with quantitatively in any way.

The provocation for the insistence in the place I have in mind is given by the word 'equality', and the context in which the word is introduced may be suggested by saying that the liberal-idealist sage and social philosopher, Sir Joshua Mattheson, who figures in *Women in Love* reminds us irresistibly of Bertrand Russell (something of a paradigmatic hero for Snow, who is himself the spiritual son of H. G. Wells). The Lawrence-like Birkin of Lawrence's novel says: 'I want every man to have his share in the world's goods, so that I am rid of his importunity . . .' The un-Lawrentian tone given by 'rid' and 'importunity' belongs to the dramatic Birkin and the dramatic context, but in what Birkin has just said we have pure Lawrence: ' "We are all different and unequal in spirit—it is only the social differences that are based on accidental material conditions. We are all abstractly and mathematically equal, if you like. Every man has hunger and thirst, two eyes, one nose and two legs. We're all the same in point of number. But spiritually, there is pure difference and neither equality nor inequality counts." '

The point is intimately related to that which Lawrence makes when he says that few people live on the spot where they are—which is equivalent to saying that few people really live. Snow, in exhorting us to put aside our individual living and live instead on 'social hope', preaches as the way of salvation the characteristic modern mode of refusing to live on the spot where one is. 'Live', of course, is a word of many possible values, as great novelists and poets make us know. Snow, refraining from permitting himself a morbid consciousness of his individual tragedy, enjoys a personal life, I suspect, that gives him considerable satisfaction—being a sage, a familiar of the Corridors of Power, a member of the Athenæum, a great figure in the Sunday papers, a great novelist, a maker of young novelists, a maker (perhaps) of academic careers. He can hardly, for the myriads for whom he generously entertains 'social hope', plan or foresee lives that will be filled with satisfaction and significance in that way. But what primarily calls for emphasis is the poverty of Snow's ostensible range of satisfactions—which is a poverty of his own canons, and of his sense of significance; a poverty in considering which one finds oneself considering the inadequacy of his sense of human nature and human need.

The significance of his blankness in the face of literature is immense. It is a significance the more damning (in relation to his pretensions) because of the conviction with which he offers himself as an authority on the literature of the present and the past. I didn't exaggerate when I said that he doesn't know what literature is. Every pronouncement he makes about it—and he makes a great many—enforces that truth. Illustrating his notion of the important kind of relation between art and life, the writer and the contemporary world, he tells us that the Russians (he knows all about Russian literature) 'are as ready to cope in art with the processes of production as Balzac was with the processes of craft manufacture'. But, for those preoccupied with the problems Snow confronts us with, unintentionally, literature has its immediate and crucial relevance because of the kind of writer who asks, who lives in his art and makes *us* live, kinds of question that, except as conventional profundities to which one should sometimes lift one's hat, seem never to have come within Snow's cognizance (an effect only emphasized by his 'tragic' and 'we die alone'—which belong, of course, to the most abject journalism). What for—what ultimately for? What, ultimately, do men live by? These questions are in and of the creative drive that produces great art in Conrad and Lawrence (to instance two very different novelists of the century who haven't, one gathers, impressed Snow).

Take, as a simple illustration, Conrad's *The Shadow Line*, and note—well, note everything, but note particularly the evocation of the young master's inner response when he first sets eyes on his ship, his first command. The urgent creative exploring represented by the questions is immeasurably more complex in *Women in Love*, a comprehensive and intensely 'engaged' study of modern civilization. Of course, to such questions there can't be, in any ordinary sense of the word, 'answers', and the effect as of total 'answer' differs as between Conrad and Lawrence, or as between any two great writers. But life in the civilization of an age for which such creative questioning is not done and is not influential on general sensibility tends characteristically to lack a dimension: it tends to have no depth—no depth against which it doesn't tacitly protect itself by the habit of unawareness (so Snow enjoins us to do our living in the dimension of 'social hope'). In coming to terms with great literature we discover what at bottom we really believe. What for—what ultimately for? What do men live by?—the questions work and tell at what I can only call a religious depth of thought and feeling. Perhaps, with my eye on the adjective, I may just recall for you Tom Brangwen, in *The Rainbow*, watching by the fold in lambing-time under the night-sky: 'He knew he did not belong to himself.'

It is characteristic of Snow that 'believe' for him should

be a very simple word. 'Statistically,' he says, 'I suppose
slightly more scientists are in religious terms unbelievers,
compared with the rest of the intellectual world.' There
are believers and unbelievers; we all know what 'religious
terms' are; and everything relevant in relation to the adjective has been said. Snow goes on at once: 'Statistically, I
suppose slightly more scientists are on the Left in open
politics.' The *naïveté* is complete; it is a *naïveté* indistin-
guishable from the portentous ignorance. The ignorance
is that which appears as historical ignorance in his account
of the Industrial Revolution, and its consequences, in the
nineteenth century. It manifests itself as a terrifying confidence of simplification—terrifying because of the dis-
tortions and falsifications it entails, and the part it plays
in that spirit of practical wisdom about the human future
of which Snow's Rede lecture might be called a classic.
Disposing with noble scorn of a wholly imaginary kind of
opposition to his crass Wellsianism, he says (and *this* is
his history—and his logic): 'For, with singular unanimity,
in any country where they have had the chance, the poor
have walked off the land into the factories as fast as the
factories could take them.' This, of course, is mere brute
assertion, callous in its irresponsibility. But it is essential
to Snow's wisdom. If one points out that the actual history
has been, with significance for one's apprehension of the
full human problem, incomparably and poignantly more
complex than that, Snow dismisses one as a 'natural Luddite'. He dismisses so—sees no further significance in—
Dickens and Ruskin, and all the writers leading down to
Lawrence. Yet—to confine myself to the non-creative
writer, about whom the challenged comment is most
easily made—it was Ruskin who put into currency the
distinction between wealth and well-being, which runs

down through Morris and the British Socialist movement to the Welfare State.

But for Ruskin 'well-being' or 'welfare' could not conceivably be matters of merely material standard of living, with the advantages of technology and scientific hygiene. And there we have the gap—the gap that is the emptiness beneath Snow's ignorance—between Snow and not only Ruskin, but the great creative writers of the century before Snow: they don't exist for him; nor does civilization. Pressing on this ancient university his sense of the urgency of the effort to which we must give ourselves, he says: 'Yet'—in spite, that is, of the 'horror' which, he says, is 'hard to look at straight'—'yet they've proved that common men can show astonishing fortitude in chasing jam tomorrow. Jam today, and men aren't at their most exciting: jam tomorrow, and one often sees them at their noblest. The transformations have also provided something which only the scientific culture can take in its stride. Yet, when we don't take it in our stride, it makes us look silly.'

The callously ugly insensitiveness of the mode of expression is wholly significant. It gives us Snow, who is wholly representative of the world, or culture, to which it belongs. It is the world in which Mr Macmillan said—or might, taking a tip from Snow, have varied his phrase by saying—'You never had so much jam'; and in which, if you are enlightened, you see that the sum of wisdom lies in expediting the processes which will ensure the Congolese, the Indonesians, the Bushmen (no, not the Bushmen—there aren't enough of them), the Chinese, the Indians, *their* increasing supplies of jam. It is the world in which the vital inspiration, the creative drive, is 'Jam tomorrow' (if you haven't any today) or (if you have it today) '*More* jam tomorrow'. It is the world in which, even at the level of the intellectual weeklies, 'standard of living' is an ultimate criterion, its raising an ultimate aim, a matter of wages and salaries and what you can buy with them, reduced hours of work, and the technological resources that make your increasing leisure worth having, so that productivity—the supremely important thing— must be kept on the rise, at whatever cost to protesting conservative habit.

Don't mistake me. I am not preaching that we should defy, or try to reverse, the accelerating movement of external civilization (the phrase sufficiently explains itself, I hope) that is determined by advancing technology. Nor am I suggesting that Snow, in so far as he is advocating improvements in scientific education, is wrong (I suspect he isn't very original). What I *am* saying is that such a concern is not enough—disastrously not enough. Snow himself is proof of that, product as he is of the initial cultural consequences of the kind of rapid change he wants to see accelerated to the utmost and assimilating all the world, bringing (he is convinced), provided we are foresighted enough to perceive that no one now will long consent to be without abundant jam, salvation and lasting felicity to all mankind.

It must be recognized, though, that he doesn't *say* 'salvation' or 'felicity', but 'jam'. And if 'jam' means (as it does) the prosperity and leisure enjoyed by our well-to-do working class, then the significant fact not noticed by Snow is that the felicity it represents cannot be regarded by a fully human mind as a matter for happy contemplation. Nor is it felt by the beneficiaries to be satisfying. I haven't time to enlarge on this last point. I will only remark that the observation is not confined to 'natural Luddites': I recently read in the *Economist* a disturbed

review of a book by a French sociologist of which the
theme is (not a new idea to us) the incapacity of the industrial worker, who—inevitably—looks on real living as
reserved for his leisure, to use his leisure in any but essentially passive ways. And this, for me, evokes that total
vision which makes Snow's 'social hope' unintoxicating
to many of us—the vision of our imminent tomorrow in
today's America: the energy, the triumphant technology,
the productivity, the high standard of living and the life-
impoverishment—the human emptiness; emptiness and
boredom craving alcohol—of one kind or another. Who
will assert that the average member of a modern society
is more fully human, or more alive, than a Bushman, an
Indian peasant, or a member of one of those poignantly
surviving primitive peoples, with their marvellous art and
skills and vital intelligence?

But I will come to the explicit positive note that has all
along been my goal (for I am not a Luddite) in this way:
the advance of science and technology means a human
future of change so rapid and of such kinds, of tests and
challenges so unprecedented, of decisions and possible
non-decisions so momentous and insidious in their consequences, that mankind—this is surely clear—will need
to be in full intelligent possession of its full humanity (and
'possession' here means, not confident ownership of that
which belongs to *us*—our property, but a basic living
deference towards that to which, opening as it does into
the unknown and itself unmeasurable, we know we belong).
I haven't chosen to say that mankind will need all its
traditional wisdom; that might suggest a kind of conserva-
tism that, so far as I am concerned, is the enemy. What we
need, and shall continue to need not less, is something with
the livingness of the deepest vital instinct; as intelligence,

a power—rooted, strong in experience, and supremely human—of creative response to the new challenges of time; something that is alien to either of Snow's cultures.

His blankness comes out when, intimating (he supposes) that his concern for university reform envisages the total educational function, he tells us how shocking it is that educated people should not be able to appreciate the Shakespeare of science. It simply hasn't occurred to him that to call the master scientific mind (say Rutherford) a Shakespeare is nothing but a cheap journalistic infelicity. He enforces his intention by telling us, after reporting the failure of his literary friends to describe the second law of thermodynamics: 'yet I was asking something which is about the equivalent of *Have you read a work of Shakespeare's?*' There *is* no scientific equivalent of that question; equations between orders so disparate are meaningless— which is not to say that the Neo-Wellsian assurance that proposes them hasn't *its* significance. More largely, Snow exclaims: 'As though the scientific edifice of the physical world were not, in its intellectual depth, complexity and articulation, the most beautiful and wonderful collective work of the mind of man.'

It is pleasant to think of Snow contemplating, daily perhaps, the intellectual depth, complexity and articulation in all their beauty. But there is a prior human achievement of collaborative creation, a more basic work of the mind of man (and more than the mind), one without which the triumphant erection of the scientific edifice would not have been possible: that is, the creation of the human world, including language. It is one we cannot rest on as something done in the past. It lives in the living creative response to change in the present. I mentioned language because it is in terms of literature that I can

most easily make my meaning plain, and because of the
answer that seems to me called for by Snow's designs on
the university. It is in the study of literature, the literature
of one's own language in the first place, that one comes to
recognize the nature and priority of the third realm (as,
unphilosophically, no doubt, I call it, talking with my
pupils), the realm of that which is neither merely private
and personal nor public in the sense that it can be brought
into the laboratory or pointed to. You cannot point to the
poem; it is 'there' only in the re-creative response of individual minds to the black marks on the page. But—a
necessary faith—it is something in which minds can meet.
The process in which this faith is justified is given fairly
enough in an account of the nature of criticism. A judgment is personal or it is nothing; you cannot take over
someone else's. The implicit form of a judgment is: This
is so, isn't it? The question is an appeal for confirmation
that the thing *is* so; implicitly that, though expecting,
characteristically, an answer in the form, 'yes, but—' the
'but' standing for qualifications, reserves, corrections.
Here we have a diagram of the collaborative-creative process in which the poem comes to be established as something 'out there', of common access in what is in some
sense a public world. It gives us, too, the nature of the
existence of English literature, a living whole that can
have its life only in the living present, in the creative
response of individuals, who collaboratively renew and
perpetuate what they participate in—a cultural community
or consciousness. More, it gives us the nature in general
of what I have called the 'third realm' to which all that
makes us human belongs.

Perhaps I need say no more by way of enforcing my
conviction that, for the sake of our humanity—our human-

ness, for the sake of a human future, we must do, with intelligent resolution and with faith, all we can to maintain the full life in the present—and life is growth—of our transmitted culture. Like Snow I look to the university. Unlike Snow, I am concerned to make it really a university, something (that is) more than a collocation of specialist departments—to make it a centre of human consciousness: perception, knowledge, judgment and responsibility. And perhaps I have sufficiently indicated on what lines I would justify my seeing the centre of a university in a vital English School. I mustn't say more now about what I mean by that, I will only say that the academic is the enemy and that the academic *can* be beaten, as we who ran *Scrutiny* for twenty years proved. We were, and knew we were, Cambridge—the essential Cambridge in spite of Cambridge: that gives you the spirit of what I have in mind. Snow gets on with what he calls 'the traditional culture' better than I do. To impress us with his anti-academic astringency, he tells us of the old Master of Jesus who said about trains running into Cambridge on Sunday: 'It is equally displeasing to God and to myself.' More to the point is that *that*, I remember, was very much the attitude of the academic powers when, thirty years ago, I wrote a pioneering book on modern poetry that made Eliot a key figure and proposed a new chart, and again when I backed Lawrence as a great writer.

It is assumed, I believe, that work in the scientific departments must be in close touch with the experimental-creative front. In the same way, for the university English School there is a creative front with which, of its function and nature, the School must be in the closest relation. I am not thinking of the fashionable idea that the right qualification for a teaching post is to be a poet—or a

63

commercially successful novelist. I am thinking again of what *Scrutiny* stood—and stands—for: of the creative work it did on the contemporary intellectual-cultural frontier in maintaining the critical function. I must not try now to say more about the way in which such a school would generate in the university a centre of consciousness (and conscience) for our civilization. I will merely insist that it is not inconceivable that Cambridge might become a place where the culture of the Sunday papers was not taken to represent the best that is thought and known in our time.

If so, it is conceivable, perhaps, that the journalistic addiction of our academic intellectuals—and journalism (in one form or another) is now the menacing disease of university 'English'—might, at Cambridge, be pretty generally recognized for the thing it is. In such a Cambridge the attention I have paid to a Snow would be unnecessary.

PREFATORY NOTE TO THE LECTURE
AS PUBLISHED IN 1962[1]

The abundant adverse comment directed against my lecture hasn't advanced the argument by leaving me something to answer. The *Spectator* was indulgent when it called the mass of correspondence it printed a 'debate'. I say 'adverse comment' because to say 'criticism' would be inappropriate: the case I presented wasn't dealt with— there was no attempt to deal with it. The angry, abusive and strikingly confident utterances of Snow's supporters merely illustrated the nature of the world or 'culture' that had made Snow a mind, a sage, and a major novelist. 'Without thinking they respond alike.' The confidence is

[1] Followed by the added part written for the American edition.

remarkable and significant because the demonstrators see themselves, unmistakably, as an intellectual *élite* and pre-eminently capable of grounded conviction, and yet, when they sense criticism by which their distinction and standing are implicitly denied, can only, with the flank-rubber's response, enact an involuntary corroboration of the criticism.

The lecture and its reception go on being referred to a great deal: there is reason, I think, for making generally accessible in print what I actually said. The issues are alive and momentous, and Sir Charles Snow's *The Two Cultures* seems likely to go on circulating—in schools and elsewhere. I have to thank Mr Yudkin for letting me print his article along with my lecture.[1] It didn't come to my notice till the lecture had been delivered and had appeared in the *Spectator*. And Mr Yudkin's criticism was wholly independent of mine: it had been published (in the *Cambridge Review*) a good while before my Richmond lecture had been thought of.

It might in a sense have been said to make my lecture unnecessary. But it puts what seems to me the unanswerable case against Snow from another approach than mine: Mr Yudkin is a research scientist (biochemist). And such a concurrence (for it is essentially that) arrived at from approaches so different seems to me to give decided point to the printing of the two critical formulations together. In my lecture, of course, my criticism of *The Two Cultures* subserves a preoccupation with a positive theme and advocacy of my own. But my argument *is*, very largely, the criticism, which is inseparable from the presentment of the positive theme. I know too, from many letters I have

---

[1] It is to be found together with *Two Cultures?—The Significance of C. P. Snow* as published separately (Chatto & Windus).

received from both sides of the Atlantic, that Snow, though widely thought of as 'public relations man' for Science, is far from being regarded with favour by all scientists. Nor had I supposed, or meant to imply, otherwise. In any case, I feel it impossible to believe that scientists in considerable numbers will not acclaim Mr Yudkin's criticism of Snow as sound—and salutary.

I have said enough by way of explaining the decision to print the two independently conceived critiques in association.

My lecture was given in England. The above paragraphs, written as a 'Prefatory Note', were addressed to a British public. I knew, however, that Snow had received much publicity in America, and that Professor Lionel Trilling, in the New York review *Commentary* (June 1962), had thought it worth while criticizing my lecture for a public that, for the most part, couldn't have read it. And when, in the above 'Note', I made my dismissing comment on my critics I wasn't intending to except Trilling: it seemed to me that he too had made no attempt to deal as a disinterested critic with what I had actually said. That is, no rejoinder was called for; he had my answer there already before him in my lecture. A general charge he brings against me is the one thing I think worth adverting to in particular here. I will say a little about it because I can at the same time indicate all there is any point in saying about another critic of my lecture whom I know to have addressed an American audience: Professor Richard Wollheim in *Partisan Review* (spring 1962).

'There can be no two opinions,' says Professor Trilling, 'about the tone in which Dr Leavis deals with Sir Charles.' More particularly the charge is that my references to

Snow's novels were gratuitous, not being necessary to my theme and argument. They are cruel in their gratuitousness, we are to gather: they are expressed, characteristically (Professor Trilling intimates), in a way calculated to cause unnecessary pain and offence. I have to comment that, in thus lending himself to the general cry that I have 'attacked' Snow (and 'attack' goes with the suggestion that I have indulged in an unpleasant display of personal animus), Professor Trilling, who passes as a vindicator of the critical function, seems to me guilty of *la trahison des clercs*. His attitude would make the essential work of the critic today impossible. It belongs to the ethos I was intent on challenging.

In my lecture I deal with certain menacing characteristics of our civilization. Snow, I start by emphasizing, is a *portent*; a portent in that, while he is in himself without distinction of any kind, so that it is ridiculous to credit him with any capacity for serious thinking about the problems on which he offers to advise the world, he has been accepted very widely in England (and in America too, I believe) as a powerful and diversely gifted mind and an authoritative voice of wisdom. If one calls attention to the clear truth about such a portent in a way that exposes to the full light of publicity the unanswerableness of the constatation, then, of course, the critical or hygienic process that achieved the result aimed at may fairly be called (what it necessarily had to be) drastic, and the portent as person may very well feel 'wounded', leaving his friends to accuse one of 'cruelty'. But those critics who call me, the perpetrator of the Richmond lecture, cruel—what excuse have they? None, I think. But the significance is clear: they have played their part in the creating of the portent, they have underwritten—at least tacitly—the

Intellect and the Sage, and they cry out with so intense an animus against the damaging constatation because its truth is so unanswerably clear. The unanswerableness is the 'cruelty' and is what has 'wounded' Snow. It would have been less 'cruel' if it had been accompanied, as it was not, by the animus that impels the intention to hurt.[1]

My lecture has no personal animus in it: the kind of drastic finality I aimed at in my dismissal of the Intellect and Sage was incompatible with that. But, of course, if by a sharpness, clarity and cogency of challenge that make it hardly possible not to see the 'cruel' truth you undo the publicity-work that has made a great public figure out of a person of undistinguished capacities, that person must inevitably feel that he has suffered an odious experience —one that he will identify with an unfeeling and destructive 'attack'. For anyone, however, who has made a name as concerned for a high intellectual standard and critical integrity to join with the 'victim's' friends and allies in passing on the public that kind of identification as a just criticism of the critic is (as I have said) no better than *trahison des clercs*.

For the supersession, in what should be the field of real intellectual and spiritual authority, of serious criteria by the power of creating publicity-values is a frightening manifestation of the way our civilization is going. It is a concomitant of the technological revolution. The kind of standards that concern the literary critic (say) can be 'there' for him to appeal to only in the existence of a public that can respond intelligently to the challenge and make its response felt. I have no doubt that there are in

---

[1] I was not warned of the editor's intention to insert cartoons into the text of my lecture as printed in the *Spectator*, and he had my indignant protest.

England today (I confine myself to speaking of my own country) the elements of such a public; there are a great many cultivated and responsible individuals, and they may be regarded as forming some sort of intellectual community. But it is not in anything like a full sense a community; the consequences of the technological revolution preclude its being a public in the way the critic needs.

Organs addressing the 'educated' public require in our time large circulations in order to satisfy the advertisement manager and subsist. The so-called 'quality' Sunday papers (the *Observer* and the *Sunday Times*), for instance, must attract and hold their million readers, and they can hope to do so only by catering at a level of appeal realistically calculated in relation to a mass public of the kind (that of the class-consciously superior middle classes, business and professional). They make a show, that is, of observing the standards of taste, education and serious intellectual interest of *haute culture*, while actually supplying 'magazine' diversion and gossip-fodder for the relaxed middle-brow. They maintain in their review pages—an aspect of this cultural phenomenon that, with my eye on my theme, I have to emphasize—the air and the reputation of performing the critical function at the highest level, their reviewers being (it is to be understood) of the intellectual élite. And indeed they are, if anything properly to be called an intellectual élite anywhere presents itself in British journalism. The critical function is performed at no higher level in the intellectual weeklies, where in fact we find not only the same kind of writer, but very largely the same names.

And here I come to an observation that must be seen, I suppose, as pointing to a marked difference between what an American with my interests and anxieties would report and what one has to reckon with in the aspect of British civilization I myself am contemplating. For America is not a tight little island, and Great Britain is. How little and how tight is brought home to one when, thinking that the preposterous and menacing absurdity of a C. P. Snow's consecrated public standing shouldn't remain longer undealt with, one challenges recognition for the patent fact that the Emperor is naked. One finds arrayed against one a comprehensive system of personal relations, the members of which (even though the use of the Christian name may not mean much[1]) know they 'belong', and observe a corresponding code.

The system has its literary-journalistic organs and foci and its institutional centres, and at the midmost, wholly in its possession on the literary side, is the BBC, with its organ *The Listener*. The BBC is an immensely potent means of giving general currency to the values of the metropolitan literary world—that which Snow calls 'the traditional culture', meaning the *élite* he supposes himself to mingle with in the *New Statesman* circle—and of getting them accepted as the 'best that is thought and known in our time'. These values can be the merest publicity-creations, created by iteration, symphonic suggestion, and the authority of the 'intellectuals' of the system, as Snow the Sage and distinguished novelist was.

Not that I suppose there weren't a number of his literary-intellectual friends who preferred not to seek opportunities to express publicly a conviction of his creative genius. For his incapacity as a novelist plainly is what, in my lecture, I say it is—total. He seems to me almost un-

---

[1] It is a significant index. Thus Dame Edith Sitwell pronounced for publication, 'Dr Leavis only attacked Charles because he is famous and writes good English.'

readable. Yet a great number of copies of his novels is to be found in every public library, and they go out a great deal. The explanation can be seen in the fact that Snow has been a major subject for lecture-directed study in WEA classes[1] and in adult education generally. And it is not only Snow's Rede lecture that is pushed for devout study on the young (our future educated class) in the upper forms of grammar schools; his novels are too. Are they not contemporary classics? And does not the British Council endorse that estimate?—the British Council, that characteristic British institution which, financed out of public funds and well-regarded by the Foreign Office, fosters the repute of British culture abroad (though not, I believe, in the United States), and, for the instruction of the world, issues on British writers (picked by itself) brochures that sell immensely at home and are to be found in every Cambridge bookshop.

A scrutiny of this British Council literature of guidance and currency-promotion will reveal that the ethos, the sense of values, the critical enlightenment it serves are those of what I have called the metropolitan literary world —the cultural world of Sir Charles Snow's 'literary intellectual'. The British Council, in fact, is another institutional centre for that world, and, like the BBC, is taken full advantage of as such. How much that is other than robust good conscience there may be in all this it is, in the nature of things, such a state of affairs having become established, difficult to say; in fact, the question hardly applies. But the system is quick to react to the threat represented by any criticism that seems to challenge its ethos; by any implicit reminder, that is, of serious standards. Such

[1] Workers' Educational Association, a democratic organization for adult education.

71

reminders cannot be tolerated; it is a matter of self-preservation. The resources of the system are deployed against the offending critic or 'influence'; if he can't be suppressed he must be, by any means, discredited.

For the 'literary world' has to maintain its sense, and the general illusion, of its own comprehensiveness. It is immune from control by the educated public—the intellectual and spiritual community in which, so far as in this technological civilization it can be effectively appealed to and can make its response felt, standards are 'there' to be evoked by the critic. The 'literary world' is its own public—the only one it, normally, is conscious of. It hates the suggestion that there might be another, one to be feared: a real educated public that doesn't take the 'literary world' seriously.

The 'literary world', in fact, with its command of all the means of publicity, virtually shuts off the educated public from effective existence. This public, in any serious sense of the noun, is only a potential public. It has no part in the formation of contemporary taste, no power to influence or to check—I am thinking of what passes for educated taste.

Though I spoke of the 'literary world' as 'metropolitan', I wasn't forgetting that, most significantly, an essential element in it (and I don't mean my 'literary' to be taken in a narrow sense—I am thinking of what, borrowing a licence from Snow, I will call the whole publicity-created culture) belongs to the universities. I refer to this fact and its significance in the close of my lecture. That Snow should have been chosen as Rede lecturer at Cambridge, that his lecture should have passed there as a distinguished intellectual performance, and that his novels should be supposed by Classical dons to be contemporary literature

—these things can't be seen as surprising: they are representative. It wouldn't be at all ridiculous to conjecture that Snow might have decisive influence in academic appointments—and on the side of the humanities.

Here, then, we have the cultural consequences of the technological revolution. And to Professor Wollheim, who charges me (so far as I can understand him) with insidious and significant evasion in not saying clearly whether I am for a high material standard of living or against it, I reply that the problem I am concerned with cannot be reduced to those terms, and that in insisting that it can he ignores the whole theme, argument and substance of my lecture, and that, if he really reads it, he will find my answer to the given charge in the place where I deal with Snow's use of 'Luddite'—and elsewhere. What I have contended, giving my reasons as forcibly as in an hour's discourse I could, is that we mustn't regard these cultural consequences as inevitable, or acquiesce in their being accepted mechanically and unconsciously, and that a pre-occupation and an effort of a very different kind from any contemplated by Snow are necessary. And I have to add that, in his ability to ignore my theme as he does, Professor Wollheim, who is a philosopher, seems to me himself a portent.

As for Professor Trilling's charge that my bringing Snow's novels into my argument was a gratuitously offensive irrelevance, I find it hard to understand how he can have persuaded himself that he was justified in making it. Is it true that, as I have heard, he has committed himself in print to a favourable opinion of Snow the novelist? If so, does he still hold it?

Professor Wollheim's suggestion ('the sheltered atmosphere of the hall at Downing College') that the Press and the BBC were excluded from my lecture because I wished to be protected against critical reactions is oddly wide of the mark. The Richmond lecture is private, and I, who was merely the person invited to give it in 1962, certainly didn't want it to be made the occasion of the ugly kind of publicity it actually got—and would inevitably get, if the Press were given a chance. And I was intent on ensuring that my actual theme and argument should be really attended to. I should have indeed been a fool if I had thought that giving the journalists an opportunity was likely to further that aim. My purpose was to see to the publication of the full text myself. It will be supposed that, in so far as I had been sanguine, Professor Trilling's and Professor Wollheim's responses to the text as it appeared in the *Spectator* brought me some disillusion.

III
Luddites?
or There is Only One Culture

LUDDITES?
or THERE IS ONLY ONE CULTURE

I AM used to being misrepresented, but not resigned to
it. Everyone who has committed himself in relation to
the themes I discussed in my Richmond lecture, *Two
Cultures?—The Significance of C. P. Snow*, and taken a line
at all like that taken by me there, knows how gross and
inconsequent is the misrepresentation that follows, and
how impossible it is to get the case one has put attended
to. Instead, something quite different is, explicitly and
implicitly, associated with one's name and made the
target for a routine play of contemptuous and dismissing
reference. Of course, this kind of response represents a
large element of willed refusal to see and understand—
the will not recognizing itself for the thing it is by reason
of a flank-rubbing consensus that is its sanction. But
this element of refusal is an essential characteristic of the
situation that the persuader has to deal with, and therefore,
if one thinks the issues are of moment—and I do, one is
not resigned.

You see, I am confessing to a touch of embarrassment:
I don't want to seem to be attributing any of that unintelli-
gent—or anti-intelligence—set of the will to the present
audience as a general characterizing trait, but, in presenting as clearly as I can in positive terms and in a positive
spirit (which is what I want to do) my view of the issues
raised by the talk about the 'two cultures', I am bound to
refer to the misrepresentations and misunderstandings that
ought not by now to need answering but seem all the same

to be the staple enlightenment about these issues so far as the publicity-practitioners, the formers of public opinion, are concerned.

And that, I must emphasize, is not merely at the lowest level. Richard Wollheim, for example, who is a Professor of Philosophy, remarked in *Partisan Review*, with an air of convicting me of insidious and significant evasion, that I had not in my Richmond lecture made clear whether I was *for* a high material standard of living or *against* it. My point of course had been this: that it won't do to make a rising material standard of living the self-sufficient aim on the confident assumption that we needn't admit any other kind of consideration, any more adequate recognition of human nature and human need, into the incitement and direction of our thinking and our effort: technological and material advance and fair distribution—it's enough, it's the only true responsibility, to concentrate on them; that's the attitude I confront.

Again, to take a very representative example of the habit of those who address the British intelligentsia, a writer in the *Spectator*, Sarah Gainham, who had earlier joined in the indignant outcry so copiously publicized in that journal against my 'attack' on Snow (which it had, at its own request, printed), ends a subsequent review of a German author with this:

> She was used to well-being, yet it is materialism for the mass of people to get used to well-being. This is a familiar resentment and envy, often seen in Britain, that working people should be going to Florence and Majorca, and buying Beethoven long-playing records. This ought not to be dressed up as moral indignation.

Of course it oughtn't, as far as it exists. And possibly the writer could back her 'familiar' by adducing instances known to her. But to suggest that such resentment and envy are representative, so that she can reasonably dismiss in this way all questioning uneasiness about the human consequences of the technological revolution and the affluent society—is that to promote clear vision and intelligent thought? The unrealism, the disturbing emotional intention, or perversity, betrays itself clearly enough in those 'Beethoven long-playing records'. I myself, after an unaffluent and very much 'engaged' academic life, am not familiar with Majorca or Florence, but in those once very quiet places very much nearer Cambridge to which my wife and I used to take our children the working-class people now everywhere to be met with in profusion carry transistors round with them almost invariably. The music that comes from these, like that one hears in greater volume in the neighbourhood of the bingo establishments (of which the smallest coast-hamlet has at least one— bingo being the most pathetic of vacuum-fillers) doesn't at all suggest aspirations towards Beethoven. If working-class people did, characteristically, or in significant numbers, show a bent that way, who would be found deploring it?—except, of course, Kingsley Amis and his admirers (and there you *have* a significant cultural phenomenon that Miss Gainham would do well to ponder). But as for the actual working-class people who *can* be regarded as characteristic, it's not anything in the nature of moral indignation one feels towards *them*, but shame, concern and apprehension at the way our civilization has let them down—left them to enjoy a 'high standard of living' in a vacuum of disinheritance. The concern, I imagine, is what all decent people capable of sympathetic perception must feel; the apprehension is for the future of humanity.

I shall hardly be accused of paradox here. It isn't very

long since the *New Statesman*—*the New Statesman*, on whose Board of Directors it was natural for Snow to be, came out with a front-page article headed 'The Menace of Leisure'. That is an irony indeed—a richly charged one. I recall a passage of D. H. Lawrence's criticism which is especially useful to those faced with enforcing the point that Lawrence was no more given to Morrisian archaizing —garden-suburb handicraftiness—than to the Carlylean gospel of Work. The passage occurs in the 'Study of Thomas Hardy' which deserves more attention than it gets, and is to be found on page 425 of *Phoenix*:

> But why so much: why repeat so often the mechanical movement? Let me not have so much of this work to do, let me not be consumed so overmuch in my own self-preservation, let me not be imprisoned in this proven, finite existence all my days.
> This has been the cry of humanity since the world began. This is the glamour of kings, the glamour of men who had the opportunity to be . . .
> Wherefore I do honour to the machine and to its inventors.

An irony! Lawrence in 1915 does honour to the machine because it gives us leisure—leisure for living (and 'living', he adds, 'is not simply not dying'), and now, for the *New Statesman*, leisure is a menace.

But my immediate point regards the way in which any writer who is known as taking a less simple view of the development and human significance of industrial civilization than Lord Snow (or Lord Robbins) is dubbed 'Luddite', after the machine-breakers, and dismissed—the implication, or contention, being that literature tends in general to be the enemy of enlightenment in this matter. The term has come significantly into favour these last two or three years. Thus Lawrence is a Luddite. And not so long ago I read in the *Sunday Times* (or the other one of those

British Sunday papers which have magazine sections in which culture gets a show) a book-notice in which the reviewer, discussing a work on the Victorian city, reeled off a list of names of distinguished Victorian writers as those of notorious Luddites in their attitude towards the new kind of urban development.

Well, though I think it serves no intelligent purpose to dub Carlyle and Ruskin 'Luddites', I could pass that with a shrug. And if Morris is dubbed 'Luddite', it doesn't move me to fierce indignation. But Arnold and Dickens! I will confine my necessarily brief commentary to Dickens, the great creative writer, for it is the dismissal of him that is most significant—I mean most revealing of the nature of the *parti pris* we have to do with in the general dismissal by the Neo-Wellsians of the thought and witness and the profoundly relevant creative energy represented by literature.

Dickens was a great novelist, and, as such, an incomparable social historian. It is the great novelists above all who give us our social history; compared with what is done in *their* work—their creative work—the histories of the professional social historian seem empty and unenlightening. Dickens himself lived intensely, experienced intensely at first hand a wide range of the life of his time, and was peculiarly well qualified to make the most of his opportunities of observing. His power of evoking contemporary reality so that it lives for us today wasn't a mere matter of vividness in rendering the surface; it went with the insight and intelligence of genius. The vitality of his art was understanding. In fact, as I have gone on reading him I have come to realize that his genius is in certain essential ways akin to Lawrence's. He saw how the diverse interplaying currents of life flowed strongly and gathered force here, dwindled there from importance to relative unimportance, settled there into something oppressively stagnant, reasserted themselves elsewhere as strong new promise. The forty years of his writing life were years of portentous change, and, in the way only a great creative writer, sensitive to the full actuality of contemporary life, could, he registers changing England in the succession of his books with wonderful vividness.

Except in so far as Coketown in *Hard Times* constitutes an exception, Dickens doesn't deal with the industrial city. The urban world of modern civilization for him is London. And it is true that he presents it as a squalid, gloomy and oppressive immensity, blighting and sinister to the life it swarms with. But to make this justify our classing Dickens as a Luddite is an odd—significantly odd—proceeding. How much less than no excuse there is for it can be brought out by recalling that in *Dombey and Son*, Dickens's first great novel—one the organizing theme of which entails a critical presentment of the contemporary civilization, the time being that of the triumphant arrival of the railway age, he symbolizes the human purpose and energy that must be looked to for an escape from the squalor, misery and confusion by the railway. There is that expedition in Chapter VI—Polly Toodle, Paul's nurse, with Paul and Florence and Susan Nipper—to Staggs's Gardens. On their way there they pass through the scene of the great earthquake that has rent Camden Town, where the new railway is being driven through to Euston terminus. The evocation of the scene is a magnificent and characteristic triumph of the Dickensian genius. As I have noted in writing about the book, we are reminded of those drawings, paintings and engravings in which the artists of that time record their sense of the Titanism and roman-

tic sublimity of the works of man. It is not merely by the
Titanic audacity, but by the human promise above all,
that Dickens is so profoundly impressed. He concludes:

> In short, the yet unfinished and unopened Railroad was in
> Progress, and from the very core of all this dire disorder trailed
> smoothly away upon its mighty course of civilization and
> improvement.

Dickens, the Luddite!—the note of this climactic sentence is not a casual inspiration, alien to the force and
feeling of *Dombey and Son*. We have a dramatic presence,
unmistakably essential to the book, and central here, in
Toodle's answer to Dombey's questioning when Polly
Toodle, his wife (the natural motherly woman, and as such
herself essential to Dickens's creative theme), is being
interviewed as the prospective wet-nurse who shall save
little Paul's life: what has Toodle's work been?—

> 'Mostly underground, Sir, till I got married. I came to the
> level then. I'm agoing on one of these here railroads, when
> they comes into full play.'

The prosperity and happiness of the Toodle family are
associated with the coming into full play of the railways,
and seen as a representative accompaniment.

By way of insisting that this characteristic of *Dombey
and Son* is characteristic of Dickens, I will just point to
Daniel Doyce, the inventor, and his place in the scheme
of values of *Little Dorrit*, that very great novel which, of
all Dickens's larger works, is the most highly organized,
everything in it being significant in relation to the whole
—and the whole constituting something like an inquest
into civilization in contemporary England. Doyce, genius
of beneficent invention, and, in the face of the Circum-
locution Office and the patronizing bourgeois (Meagles),
invincibly sane, persistent and matter-of-fact, pairs with

83

Cavalletto, the little Italian who puzzles the inhabitants of Bleeding Heart Yard by his simple ability to live and enjoy the sun: neither a major actor, they are major presences for the dramatic and poetic process of valuation implicit in Dickens's art because of what they so potently are and represent.

What I have been trying to bring out for clear recognition is the element of deep-seated refusal to perceive that betrays itself in such characteristic instances as I have adduced. We have it not only in the dead set at eliminating literature and what it represents from all serious relevance to the issues, but in the attitude towards *any* suggestion that the issues are essentially more complex than Snow's Rede lecture would make them, and fraught with other kinds of menace to humanity than he is able to recognize. There is that business of 'the old wheelwright's shop'— the play made with that phrase, thrown out with a knowing 'ha-ha' in the voice and manner, by a well-known Cambridge figure of the BBC world. The ironist is what Snow calls a 'literary intellectual', and the blind set of the will he relies on and means to confirm is that which had a representative illustration when a bright lady journalist in the *Spectator* quite wantonly dragged my name in as that (everyone knew) of the man to apply to if you wanted a wistful lament for the Old Style Pub (now vanished, I gather—for actually I know nothing about these things). The sole factual basis that could be alleged for these insinuations is the use to which, thirty years ago and more, Denys Thompson and myself in *Culture and Environment*, a book for schools, put George Sturt's *The Wheelwright's Shop*.

The use to which we put Sturt had nothing William-Morrisian in it; neither of us, I may say, went in for folk-dancing—or pubs. The attention we aimed at promoting was to the present, and our emphasis was on the need to understand the nature of the accelerating and inevitable change that was transforming our civilization. The wheelwright's business, we pointed out, or noted how Sturt pointed out, didn't merely provide him with a satisfying craft that entailed the use of a diversity of skills; it contained a full human meaning in itself—it kept a human significance always present, and this was a climate in which the craftsman lived and worked: lived *as* he worked. Its materials were for the most part locally grown, and the wheelwright quite commonly had noted as a tree *in situ* the timber that came to the shop—which is a representative aspect of the general truth. The customers too were local, and he knew them, themselves and their settings, as meeting their particular requirements he had to, individually—he, the wheelwright of the neighbourhood. He saw the products of his craft in use, serving their functions in the life and purpose of a community that really *was* a community, a human microcosm, and couldn't help feeling itself one.

We didn't recall this organic kind of relation of work to life in any nostalgic spirit, as something to be restored, or to take a melancholy pleasure in lamenting; but by way of emphasizing that it was *gone*, with the organic community it belonged to, not to be restored in any foreseeable future. We were calling attention to an essential change in human conditions that is entailed by the accelerating technological revolution, and to the nature of the attendant human problem. And our sense that the problem was not likely to get all the attention it should be seen as demanding has been redundantly justified.

It is plain that the kind of relation between work and

living documented in *The Wheelwright's Shop*, or anything like it, can't by any serious mind be proposed as an
ideal aim in *our* world; that the only bearing it has on the
possibilities we have to consider is that to recognize the
nature of the change is to recognize the nature of the
challenge, the problem that Snow ignores, a frightening
characteristic that it has being to escape notice for what
it is.

Mr Toodle of *Dombey and Son*, as stoker and aspiring
engine-driver, had a job out of which he got much satisfaction, besides that of being able to support his family.
He had there, we know, an advantage over the mass of
industrial workers, and we have no difficulty in understanding how Dickens could present him as invested with
a cheering significance for the human future. But the
future we now see for the Toodleses is automation, and
the future seen as automation is what makes the *New
Statesman* talk of the 'menace of leisure'. The development by which, for industrial workers, real living tends
to be something thought of as saved for the leisure part of
life is soon to be consummated. The meaninglessness, or
human emptiness, of work will be sufferable because the
working part of life will be comparatively short and the
leisure part preponderant.

That this upshot of technological progress needs to be
thought of as facing us with a problem is, as I've noted,
receiving some kind of recognition: 'the menace of leisure',
'education for leisure', and so on—phrases of that kind
give us the nature of the recognition. And my point is
that such recognition is no real recognition of the problem
that faces humanity. Certainly that problem is not being
recognized for what it really is when discussion proceeds
in terms of the need to educate for positive and more satis-

fying uses of his leisure the worker whose routine work, requiring or permitting no creative effort on his part, and no large active interest—little more, in fact, than automatisms—leaves him incapable of any but the passive and the crude.

We are *all* involved—and in the most intimate, inward and essential way; and not merely by reason of congested roads, the smell of fish and chips, the ubiquity of transistors and that kind of inconvenience, which is what, in England, discussion pertaining to the ethos of the *New Statesman* tends, however democratically (of course), to suggest. A general impoverishment of life—that is the threat that, ironically, accompanies the technological advance and the rising standard of living; and we are all involved.

Snow gives us a pregnant demonstration, very pertinent to the explaining of what I mean, when, as himself representative of his Two Cultures, he posits, to be set over against what he calls the 'scientific culture', a 'literary culture' that he represents by the literary intellectual of the *New Statesman* milieu, or the modish London literary world. Well, our traditional culture hasn't yet been finally reduced, though Snow in his Rede lecture reduces it, to *that*. But his being able to do it quite naturally, without a thought of being questioned, shows where we have got. And the process so exemplified—I permit myself to say what should be obvious—affects the scientist, the scientist as a man, as nearly and intimately as anyone.

I say it, because no one will suggest that *he*, the real scientist (or the technological expert, for that matter), is to be thought of as sharing the state of the human quasi-automaton—the human or animal or organic adjunct to automation. The scientist very well may—the creative kind certainly will—derive great satisfaction from his work. But he cannot derive from it all that a human being needs—intellectually, spiritually, culturally. Yet to think of a distinguished mind having to go for refreshment, edification and nourishment to the 'literary culture' represented by Snow's 'literary intellectual' and identified by Snow with 'the traditional culture' is painful and depressing—unbearably humiliating to some of us. For I was speaking responsibly when I said in my Richmond lecture that Snow's 'literary intellectual' is an enemy of art and life. He belongs to the cultural conditions that make it seem plausible—obvious good sense—to talk about 'The Two Cultures'.

The term 'culture', of course, like most important words, has more forces than one in which it can be used for intellectually respectable purposes; but even if Snow had not with naïve explicitness identified one of his pair with 'the traditional culture', the fair and final dismissing comment on his Rede lecture as offering serious thought about the problem he points to (but doesn't see) would be: 'there is only *one* culture; to talk of *two* in your way is to use an essential term with obviously disqualifying irresponsibility'. It is obviously absurd to posit a 'culture' that the scientist has *qua* scientist. What Snow proposes to condemn the scientist to when he points to the really educated man as combining the 'two cultures', what he condemns the scientist to for his cultural needs—his non-scientific *human* needs, is (in the British terms I am familiar with) the culture of the *New Statesman* and the Sunday papers, which is what Snow's 'literary intellectual' actually represents. And for that I have intimated my contempt. No serious problem or theme is being tackled when an alleged 'scientific culture' is being placed against *that* as the complementary reality.

The difficulty about proceeding now on a more positive line is that the issues, being basic, are so large, complex and difficult to limit, and that (a distinctive mark of the present phase of civilization) even in talking to a highly educated audience there is so little one can take as given and granted and understood to be necessarily granted. One can only be clear about one's focal interest and determine one's course and one's economy in relation to that. Mine—ours, may I say?—is the university; that is, the function and the idea.

It may be commented at this point that I am not absolved from explaining what positively I mean by 'culture' in the sense I invoke when I criticize misleading uses of the term. I won't proceed by attempting to offer a direct formal definition; that wouldn't conduce to economy, or to the kind of clarity that for the present purpose we need. Faced with the problem of indicating clearly the nature of my answer to questions about meaning, I recall Snow's account of the supremely creative human achievement present to us in Science, and the comment it moved me to. This is Snow: 'As though the scientific edifice of the physical world', he exclaims, 'were not, in its intellectual depth, complexity and articulation, the most beautiful and wonderful collective work of the mind of man.' My comment was: 'It is pleasant to think of Snow contemplating, daily perhaps, the intellectual depth, complexity and articulation in all their beauty. But there is a prior human achievement of collaborative creation, a more basic work of the mind of man (and more than the mind), one without which the triumphant erection of the scientific edifice would not have been possible: that is, the creation of the human world, including language.'

This is surely a clear enough truth, and I don't suppose

anyone here wants to dispute it. The trouble is that in our time, when we need as no other age did before to see that it is given full conscious realizing recognition, there seems to be something like an impossibility of getting anything better than a mere notional assent. Can we without exaggeration say, for instance, that it was even that in the *Guardian* first leader which I read when, a couple of years ago, my thoughts were very much on the theme I am discussing now? The leader, though characteristic enough, a little surprised me, for I had had strong grounds for supposing the *Guardian* pro-Snow and anti-Leavis—committed to the view that there was nothing to be said for my side in the notorious so-called 'debate'. But this leader, dealing with Mr Harold Wilson's 'vision of the future' at Scarborough, remarked that 'even a C. P. Snow would be a poor substitute for an informed and open discussion of the uses to be made of science', and that it 'would . . . be a tragedy if either party gave as its chief reason for a further extension of the universities the need to recruit more scientists'. It concluded: 'Science is a means to an end.'

'Science is a means to an end': what more could one ask?—it concedes everything. You'll reply that 'concedes' is an infelicitous word, the proposition being a truism. Yes, a truism: there's the rub. But 'rub' itself is an infelicitous word: it doesn't—quite the contrary—suggest what I am calling attention to, which is the absence, where 'ends' are adverted to as needing some consideration, of the friction, the sense of pregnant arrest, which goes with active realizing thought and the taking of a real charged meaning. 'Science is a means to an end': yes— a rising standard of living. I perpetrate my notorious exhibition of bad manners at poor Lord Snow's expense, or, as I myself should put it, do my best to get some recognition for the inadequacy of that accepted formula as representing a due concern for human ends (a matter, it seems to me, of great urgency), and I get for response a vast deal of blackguarding, misrepresentation and contemptuous dismissal, and then the *Spectator* offering to strike a *juste milieu* between my Richmond lecture on the one hand and Aldous Huxley offering to strike a *juste milieu* between me and Snow on the other. The *Spectator*, pointing out how unsatisfactory Huxley is, endorses by making it its own his attribution to me of 'one-track, moralistic literarism'. To set over against Snow's deviation, scientism, you see, there is mine, which is literarism.

I'll leave aside for a moment this curious term, which Huxley, with an American distribution of stress and quantity (after all he, though the *Spectator* challenges for itself with the term the genuine and solid middle position it was meant to claim for him, invented it) found perhaps more speakable than I do. Immediately in place is to insist on the truth—not, unhappily, a truism—that once the naïvety that takes 'rising standard of living' to represent an adequate concern for human ends has been transcended, the determination of what, adequately conceived, they *are* is seen to be very far from simple. Human nature and need are certainly more complex than Lord Snow assumes. They won't be fully apparent for recognition in any present of any society. The most carefully analysed and interpreted answers to the most cunningly framed questionnaires, the most searching and thorough sociological surveys, won't yield an adequate account of what they are.

And 'end' itself, though a word that we most certainly have to use in the kind of context and the kind of way I've been exemplifying, tends, perhaps, to turn the receptivity of the mind away from orders of consideration that are essential—essential, that is, when the criteria for determining how we should discriminate and judge in the face of a rapidly changing civilization are what we want to bring to full consciousness. Mankind, for instance, has a need to feel life significant; a hunger for significance that isn't altogether satisfied by devotion to Tottenham Hotspur or by hopes of the World Cup for a team called England or Uruguay, or by space travel (mediated by professional publicists), or by patriotic ardour nourished on international athletics, or by the thrill of broken records —even though records, by dint of scientific training, go on being broken and the measurement of times becomes progressively finer.

If, of course, one is challenged to stand and deliver and say what 'significance' is—'If you use the term you ought to be able to say what you mean by it!'—it is hardly possible to answer convincingly at the level of the challenge. But that is far from saying that the matter for consideration raised with the term is not, when thought turns on human ends, of the greatest moment. And that 'high standard of living' expresses a dangerously inadequate notion or criterion of human prosperity is a simple enough truth.

Now, if we are asked how we are to arrive—for ourselves, in the first place, but of course, not merely that— at a more adequate notion, the answer, it seems to me, clearly is: when human ends require to be pondered in relation to the pressing problems and opportunities with which our civilization faces us, one's thinking should not be blind to the insights given in cultural tradition—on the contrary, it should be informed with the knowledge of basic human need that is transmitted by *that*. This is not a simple answer; no serious answer *could* be. I have used the phrase 'cultural tradition' rather than Snow's 'the traditional culture', because this last suggests something quite different from what I mean. It suggests something belonging to the past, a reservoir of alleged wisdom, an established habit, an unadventurousness in the face of life and change. Let me, as against that, extend briefly the quotation I've permitted myself from my Richmond lecture. Having, in comment on Snow's claim regarding the 'scientific edifice of the physical world', pointed to the 'prior human achievement of collaborative creation ... the creation of the human world, including language', I go on: 'It is one we cannot rest on as on something done in the past. It lives in the living creative response to change in the present.' A little further on, insisting on the antithesis to what 'traditional' usually suggests, I put it in this way, and the formulation gives me what I need now: 'for the sake of our humanity—our humanness, for the sake of a human future we must do, with intelligent resolution and with faith, all we can to maintain the full life in the present —and life is growth—of our transmitted culture'.

We have no other; there is only one, and there can be no substitute. Those who talk of two and of joining them would present us impressively with the sum of two nothings: it is the void the modern world tackles with drugs, sex and alcohol.[1] That kind of sage doesn't touch on the real problem; he has no cognizance of it. It is a desperately difficult problem; I don't pretend to know of comfortable answers and easy solutions. Simply I believe that in respect of this problem, too, intelligence-directed human effort has *its* part to play, and that there is a human instinct of self-preservation to be appealed to.

[1] And, I can now add, 'student unrest' and the vote and majority-status at eighteen.

In my Richmond lecture, recalling a formulation I had been prompted to in the old days, when the Marxising expositors of human affairs thronged the arena, I remarked that there is a certain autonomy of the human spirit. I didn't mean by that to suggest that the higher non-material achievements of human culture, the achievements of collaborative creation that belong most obviously to what I call in discussion the 'third realm', were to be thought of as spontaneous, unconditioned expressions of an intrinsic human nature sprouting or creating gratuitously, in a realm of pure spirit. I was merely insisting that there *is* an intrinsic human nature, with needs and latent potentialities the most brilliant scientist may very well be blank about, and the technologically-directed planner may ignore—with (it doesn't need arguing) disastrous consequences. Of course, the collaborative creation of the world of significances and values has to be seen as a matter of response to material conditions and economic necessities.

Let me repeat, however, that I didn't thirty years ago point to the state of affairs, the relation between cultural values (or—shall I say?—human significances) and economic fact, documented in *The Wheelwright's Shop*—which I've hardly mentioned these thirty years—as something we should aim at recovering; but as something finally gone. That relation was an essential condition of the kind of achievement of the higher culture (spiritual, intellectual, humane) that is represented by Shakespeare's works. Such a relation, for any world we can foresee, is gone.

Technological change has marked cultural consequences. There is an implicit logic that will impose, if not met by creative intelligence and corrective purpose, simplifying and reductive criteria of human need and human

good, and generate, to form the mind and spirit of civilization, disastrously false and inadequate conceptions of the
ends to which science should be a means. This logic or
drive is immensely and insidiously powerful. Its tendency
appears very plainly in the cultural effects of mass-
production—in the levelling-down that goes with stan-
dardization. Ponder, I find myself saying in England, to
academic audiences, the 'magazine sections' of the Sunday
papers (they know which two I mean), and tell yourselves
that *this*, for many dons—and I am thinking of the non-
scientists, the custodians of culture—represents the top
level: what Arnold meant by 'the best that is thought and
known in our time'. It will almost certainly represent the
top level for those who, at this time of rapid and confident
and large scale reforms, make the authoritative and decisive
recommendations in the field of higher education.

To point out these things is not to be a Luddite. It is
to insist on the truth that, in an age of revolutionary and
constantly advancing technology, the sustained collaborative devotion of directed energy and directing intelligence
that is science needs to be accompanied by another, and
quite different, devotion of purpose and energy, another
sustained collaborative effort of creative intelligence. I
will again quote what I actually said in the offending lecture: 'the advance of science and technology means a
human future of change so rapid and of such kinds, of
tests and challenges so unprecedented, of decisions and
possible non-decisions so momentous and insidious in their
consequences, that mankind—this is surely clear—will
need to be in full intelligent possession of its full humanity. . . . What we need, and shall continue to need not
less, is something with the livingness of the deepest vital
instinct; as intelligence, a power—rooted, strong in

95

experience, and supremely human—of creative response to the new challenges of time; something that is alien to both of Snow's cultures.'

What I have been pointing out is that we shall not have this power if provision is not made for a more conscious and deliberate fostering of it than has characterized our civilization in the past. And here comes in my concern for the idea of the university as a focus of consciousness and human responsibility.

I must close on the note of transition, for I can't here follow up this opening. My mention of the idea of the university is a concluding emphasis on the positive; an insistence that my attitude is very much a positive one, and that I *have* a positive theme for the development of which I am fully charged—a theme intent on practice. Of course, I speak—have been speaking (that was plain—and was expected of me)—as an Englishman, and my 'engaged' preoccupation with the idea of the university has a British context, the tight little island of Mr Harold Wilson's premiership and Lord Robbins's Report on Higher Education. Yet I can't think of the differences between the situation I face at home in England and the situation in America (which, as I've remarked, is not a tight little island) without telling myself with conviction that we face in essence one and the same problem, and that the widest community of the intelligently concerned that can be made aware of itself and of the menace will not be too large. I have reason for knowing, with encouragement and gratitude, that there are many Americans who feel the same.

The special bent of my positive concern is given when I gloss 'the university as a focus of consciousness and human responsibility' by 'the university as a guarantor of a real performance of the critical function—that critical function which is a creative one'. It is here, of course, that I am supposed to have laid myself open to the charge of 'literarism'—for it is obviously meant to be a charge. Its suggestion is very much that of the writer in the *Melbourne Quarterly* who, in a quite flattering article, said I sometimes seem to think that literary criticism will save us. The possibility of this irony meaning anything seems to depend on a conception of literary criticism that, when I try, I can't conceive—it eludes me.

For—and this is my reply to Aldous Huxley as well—I don't believe in any 'literary values', and you won't find me talking about them; the judgments the literary critic is concerned with are judgments about life. What the critical discipline is concerned with is relevance and precision in making and developing them. To think that to have a vital contemporary performance of the critical function matters is to think that creative literature matters; and it matters because to have a living literature, a literary tradition that *lives* in the present—and nothing lives unless it goes on being creative, is to have, as an informing spirit in civilization, an informed, charged and authoritative awareness of inner human nature and human need.

In my discussions of the university English School as a liaison centre I have been intent on enforcing my conviction as to the kind of effort by which we must promote the growth of that power of which I have just spoken: 'a power—rooted, strong in experience and supremely human—of creative response to the new challenges of time; something that is alien to both of Snow's cultures'. My concern for the idea of an English School isn't to be thought of as just a matter of syllabus, teaching methods and a given kind of student product to be turned out. The educational problem itself in that narrow sense conduces to discouragement, despair and cynicism when approached merely in those terms.

But I won't now develop that observation beyond saying that one's essential concern should conceive itself as being to make the university what it ought to be—something (that is) more than a collocation of specialist departments: a centre of consciousness for the community. The problem is to re-establish an effective educated public, for it is only in the existence of an educated public, capable of responding and making its response felt, that 'standards' can be there for the critic to appeal to. This is true not merely of literary criticism; the literary-critical judgment is the type of all judgments and valuations belonging to what in my unphilosophical way I've formed the habit of calling the 'third realm'—the collaboratively created human world, the realm of what is *neither* public in the sense belonging to science (it can't be weighed or tripped over or brought into the laboratory or pointed to) *nor* merely private and personal (consider the nature of a language, of the language we can't do without—and literature is a manifestation of language). One's aim is that the university itself, having a real and vital centre of consciousness, should *be* such a public or community as the critic needs, being in that way one of the sustaining creative nuclei of a larger community.

One might then hope that one might, at Cambridge, for example—my own university (and Lord Snow's and Lord Annan's)—get the effective response when one uttered the appropriate judgment on the publicizers, public relations men, heads of houses, academic ward-bosses, hobnobbers with Cabinet ministers, who are planning, they tell us, to remodel the University and start going a new kind of higher education. Might . . . might, if . . . if . . .; but don't take me to be suggesting that the actuality and the blind enlightened menace are anything but what they are. Let me end with a sentence and a bit from Lord Robbins of the Robbins Report:

> Since Sir Charles Snow's Rede lecture, we have heard a great deal of the two cultures in this country; for reasons which I completely fail to understand, Sir Charles's very moderate indication of danger arouses very high passions. To me his diagnosis seems obvious . . .

To me it seems obvious that Snow's *raison d'être* is to be an elementary test.

IV
'*English*', *Unrest and Continuity*

'ENGLISH', UNREST AND CONTINUITY

THERE could be in our time no more important pre-occupation than that which brings us here. To say that isn't mere departmental parochialism; it's to recognize that I've undertaken something very difficult. You can't discuss intelligently how we should conceive university 'English' except in relation to an idea of the university, its place and function in our unprecedented world. You can't be intelligent about what the university should be except by being intelligent about the notorious social disorders everyone reads about in the papers, and the menaces, not so notorious, with which our civilization confronts (but doesn't suitably frighten) mankind. As a way both of imparting my sense of the difficulty and of facilitating economies I'll read a newspaper cutting—actually it's a letter I wrote to *The Times* not long ago, when the kind of situation I must give some attention to was already very much in my mind.

Sir,
    In your leading article, 'The British Backlash', there is one thing that is not said, and that the most important. In its absence what you do say gives ground only for despair. 'Student revolt' can't be intelligently discussed as an isolated thing. Popular resentment generated by student violence and fatuity may lead to greater firmness in the universities (and I hope there will be firmness), but that won't itself cure the disease there; and elsewhere all the familiar manifestations—violence, wanton destructiveness, the drug menace, adolescent promiscuity, permissiveness, the enlightened praise of the young for

their 'candour' about sex—will go on unabated. To isolate 'student revolt' is to promote blankness about the nature of the disease—and blankness is a major manifestation.

There is no simple remedy and no possibility of a rapid cure. But your leading article might properly have pointed out that in default of a more positive and intelligent conception of the university and its rôle in civilization than is entertained by (say) Lord Robbins and Lord Annan, or any influential politician who has made his views known, not only is there no hope of ending 'student unrest', there is no hope that the first of the three problems listed by Mr Cecil King several columns across the page from your leading article will begin to be tackled. It is (he says) the problem 'how to establish a moral authority at the centre which will give our young people leadership and contain the rising tide of crime'.

That is not an adequate conception of the problem, which is one of cultural disinheritance and the meaninglessness of the technologico-Benthamite world: to state it as Mr Cecil King does amounts to saying that nothing can be done. And in fact nothing will be done unless society commits itself to a sustained creative effort of a new kind—the effort to re-establish an educated, well-informed, responsible and influential public—a public that statesmen, administrators, editors and newspaper proprietors can respect and rely on as well as fear. Society's only conceivable organ for such an effort is the university, conceived as a creative centre of civilization.

My focal concern in writing that was ours, the English School, which, you'll have noted, gets no mention. I felt that something ought to be said, the opportunity made something of, and some telling emphasis registered that should have a clear *ad hoc* relevance to the occasion—the co-presence of *The Times* leader with Mr Cecil King's pronouncement on the alarming portents of the age and their cure: hence my letter. For you it is only a reminder of what is perhaps the most formidable face of the problem we confront—the impossibility (even if given much more space than a printable letter) of stating it in a newspaper, or to a politician (and all statesmen are politicians), or to a committee presided over by Lord Annan or Lord Robbins, so as to get it intelligently attended to. Even in talking to you, I find its depth and the kind of comprehensiveness—the reaching and branching out—that make it peculiarly complex for profitable discussion intimidating. How, in an hour or so, can I hope to do what would be needed to ensure that the particular approaches, moves, questions and thrusts that discussion *is* engage a sufficient context? But I conclude from your having honoured me with the invitation to open this discussion that, knowing me to be interested in the theme, you know also what the general nature of my views—what my own considered approach—is. That helps: it licenses an elliptical economy on my part; it frees me for a tactical flexibility—I needn't make a show of sustained expository method.

I shan't, then, start by giving you a prefatory account of the sickness of civilization. I will turn at once to the centre—our centre, the focus of preoccupation for us, which is the university English School. I know I can't assume that everyone who holds a post in 'English' believes intensely that English Literature matters—believes that it ought to be a potent living reality in the present, so that to succeed in making it that would be to do something important towards remedying those disorders of civilized society which frighten us. I can, though, assume such a belief as common to us all here. But a peculiar condition of our time is blankness, and it's difficult not to catch some infection of it. We therefore, assembled here to discuss our responsibilities and clarify our sense of them, have to cultivate consciousness, that full and sure consciousness which manifests itself in an ability to be articulate and

cogent about our assumptions and aims and the grounds
for them, as about the methods by which the aims are
to be pursued. The drive and insidious suggestion of the
age are against what we stand for—which is a way of
saying that what we stand for is what the age desperately
needs. For ourselves as well as for the blankly or blandly
questioning we have to be ready with the answers, even
if the totality of the truth that, for us, these engage is un-
likely to compel immediate recognition. I am thinking,
for instance, of the profundity of solemn doubt mimed by
Mr George Steiner and Professor Daiches (representative
intellectuals, after all) when they suggest that, contemplating Auschwitz and Dachau and the fact that finely
cultured persons countenanced these horrors, a judicious
mind sees reason for questioning the value attributed by
Leavis to the study of literature.

One can reply at once that this kind of comment takes
no notice either of any views or proposals actually advanced or of the problems to which they are addressed,
and that, whatever 'cultured'—or 'cultivated'—may mean,
no one contemplating the problems seriously would talk,
or think, of the appropriate 'solutions' or remedial recourses as a mere matter of arranging to increase the
number of cultured individuals. But this is only negative.
What positively does one reply? What aims, claims and
conceptions justify the importance we ascribe to 'English'?
We need to be able to answer. The representative quality
of the pair I have cited is something we can't ignore: one
of them, we may reflect, holds a chair in the Humanities,
and both are distinguished minds in the world of the
Guardian, the *New Statesman*, and the Sunday magazine-
sections. That, the world of enlightenment, is the great
enemy we have to fight, and to go on fighting, in what

from our side is the *creative* battle. There is no Stalingrad to be achieved over that enemy, but we can discredit its clichés, disturb its blank incuria, and undermine its assurance—things which most certainly have to be done, and will only be done if the creative purpose in us is strongly and articulately conscious (which is to say, energized by a realization of what's involved).

Well, here's a cliché—I took this instance from a letter in *The Times* for December 21: 'University work falls into two main categories—contributions to knowledge and communicating knowledge to students.' I might, as you know, have found it in many other places; we hear it whenever the nature of a university is being discussed. It is produced and accepted as a basic axiom, and we must aim not only at discrediting it, but at driving it out of the field of respectable discourse. If we countenance it even tacitly we endorse Lord Robbins's attitude towards the study of literature—his way of recognizing that English, the model of a university being under consideration, has claims to a place. English comes under the Arts, and the Arts belong to that margin which (the usefulness of the word 'aesthetic' makes itself felt here) we allocate to the graces of life—to the higher amenities from which we derive (it's proper to recognize) a sense of its dignity. They deserve attention and shouldn't be forgotten. Lord Robbins makes the significance of his attitude perfectly plain. Pointing out that the advancement of those sciences on which our technological civilization so obviously depends is not the sole business of a university, and that there clearly needs to be provision too for advancing and communicating the available knowledge of human nature and society, he emphasizes the supremely relevant importance of psychology and the Social Studies.

Here, of course, without suspecting it (that's what's so frightening), he drops a cue for us to take up. But to complete my own emphasis: if we don't, giving with conviction the strong positive grounds—which means presenting (and serving in action) our own very different conception of a university, explicitly repudiate Lord Robbins's, we shall have resigned ourselves to accepting the idea of English as a soft option, with all the futility that involves. Lord Robbins has the world with him, and whatever resistance professional pride and the sense of responsibility may offer, the pressures and the insidious climate will prevail if the pride and the responsibility are not those of a completely and strongly conceived function.

I find myself from time to time saying—some here may have heard me say—that we have to insist, and compel for English, the corresponding respect (there is no respect as things are), that we stand for a discipline of intelligence as genuine as that of any of the sciences and certainly not less important. It was that tactical emphasis I had in mind when, writing an essay on *Anna Karenina*, I chose (beforehand) for subtitle: 'Thought and Significance in a Great Creative Work'. The discipline is *sui generis*. But to insist effectively on these things implies some sort of apprehension on the part of the enemy of a whole context that gives them their meaning and force. So to turn now to the cliché about the 'two categories' under which university work falls—'contribution to knowledge and communicating knowledge to students': by it English is excluded from among serious studies. Neither the distinction nor the terminology, the language, applies where English is concerned. When enforcing this assertion positively one is insisting both on the unique nature of English as a university study and on its central and basic importance. My aver-

sion from the word 'teach' preceded—it was only in-
tensified by—my acquaintance with those American 'notes
on contributors' that tell one that X or Y or Z 'teaches'
Joyce or Thomas Mann or Faulkner or the *Cantos* at this
or the other college or university. If one's concern is essentially with literature one doesn't think of oneself as
'teaching'. One thinks of oneself as engaged with one's
students in the business of criticism—which, of its nature,
is collaborative. The student, on his part, ought to be able
to think of himself as belonging to a collaborative community formed by the English School as a whole, undergraduates, graduate students and permanencies, and the
more, let me add parenthetically, he can feel that it transcends departmental frontiers the better, the community
being a model or paradigm of the ideal—for it doesn't
now exist—educated public that (ideally) makes possible
at any time a performance of the function of criticism. The
collaboration is essentially a creative one.

And I don't contradict that when I say that the community should see itself as essentially aiming in its work
—in its total collaboration—at performing the function
of criticism in our time. This proposition isn't heroic
crusading elevation, but practical good sense and absolutely necessary realism *vis-à-vis* the bi-categorists and
Robbinses. I don't need to argue that there is no acquisition, no taking possession, of creative works without
value-judgment, and that judgments can't be taken over;
they are made in genuine personal self-commitment by
each student for himself, or there is no judging, and no
acquisition, so far as he is concerned. Nor need I elaborate
the case about the nature of judgment that some here will
feel, perhaps, I have insisted on too much. I will confine
myself to a minimal reminder. The form of a judgment

being 'This is so, isn't it?', the question is a request for corroboration; but it is prepared for an answer in the form 'Yes, but——', the 'but' standing for qualifications, reservations, additions, corrections. And here we have the paradigm of the process by which the poem (say) is established as something of common access standing 'out there' in what is in some sense a public world. We have at the same time a pregnant hint of the nature of the Third Realm and the way in which it is creatively renewed and kept in continuous being. In my philosophical innocence I hit, let me explain, on this term as a way of laying an emphasis on that creative human reality of significances, values and non-mensurable ends which our technologico-Benthamite civilization ignores and progressively impoverishes, thus threatening human existence. It is this blankness (and it manifests itself a great deal, both among the masses and the enlightened élite, as hostility) that I have in mind when I speak of 'spiritual Philistinism'.

Of course, we shan't make much impression on the bi-categorists, Annans and Robbinses——on either the conscientious or the unconscientious promoters of spiritual Philistinism——by talk about the nature of the existence of a poem. Indeed, it wouldn't be realism to count on making a decisive impression on them even by such an address as I hope to have given by the time I stop. Great changes are not effected so easily. When I spoke of realism I was thinking of that truth, as well as of the need to issue a clear challenge to the promoters of spiritual Philistinism at once——a challenge that will be generally recognized to be one (for that is immensely desirable). To try again; I was thinking of the obvious truth that, in such a world as that in which we live, it is only by cultivating the fullest understanding of what our responsibility to that world is that we shall conquer footing and licence to make a serious attempt at discharging it.

No one, it need hardly be said, with professional responsibility in a Department of English would think of an English course as a matter of some poems, or a selection of novels, or even an assortment of authors, to be studied. We all here know that as guides and fellow-students of our students we are concerned with relations between works, between the creative achievements of different authors, between different pasts, and between the past and the present. These matters of consideration are generally thought of as coming under the head of Literary History. Let me say at once that I'm not at all happy about what commonly comes under that head, or about what's commonly understood by Literary History —'understood', as a matter of fact, seems to be a misleading word. Indeed, I think that the question, What *is* Literary History?, is one that all the members, senior and junior, of an English School ought to have steadily in mind. But I don't think they will be making what they ought of it if it doesn't bring them constantly back to another question, which is much more fundamental: What *is* English Literature?

It is a question to which an English School should be constantly working out the answer—not a theoretical, but a concrete answer. This is to formulate again the proposition that I threw out earlier: that an English School should conceive its business as being to perform, or to make a serious attempt at performing, the function of criticism in our time. For English Literature has its life in the present, or not at all. It will be 'there', if it is to be a potent living reality, in a public that cares—and cares intelligently—about it; and that it shall be a living reality it is the function of criticism to ensure. The performance of that function implies a collaborative interplay, so that in a state of cultural health there would be more than one intelligent critic practising—there would be a whole corps of them (and there would be an expectation in an intelligent public of serious current criticism in the weekly journals and the newspapers). But the function of criticism will be far from fully performed if there is no critic writing who is not original in a major way—and 'original' here means capable of the innovating criticism that, however strong the resistance, can get its essential judgments accepted by reason of their manifest irresistibleness. Only such a critic, in an age when a notable change in cultivated taste, predisposition and critical assumption is called for, can make the necessary judgments about the present out of a profound and vivid sense of the relation of the present to the past.

I am not counting on the presence of such a critic in every English School—or in any. The point I've wanted to enforce is that it would be absurd not to make the most of T. S. Eliot. For if Eliot (here's my immediate emphasis) is a decisively original critic, and a critic of whom it may still be said that he belongs to our time, the discussion of his value in relation to my argument entails a critical appreciation of his whole achievement. I am a little embarrassed by my consciousness that I have twice already in public disquisitions invoked Eliot in this way—at Cheltenham recently and two years ago in my Clark lectures. But that fact, after all, only testifies to my confirmed conviction that the absurdity of not giving him to the full the attention he so obviously challenges will be recognized by everyone who confronts our problem. More bluntly, he's there, and we can't do without him. He serves incomparably well the need to introduce into the student's work from the beginning the organic structural principle or impulsion (in the form, necessarily, of perceptions and apprehensions that are energies, and impel to growth and living organization).

I will revert here for a moment to the head of realism. I have found myself from time to time commenting at Cambridge on the monstrous *un*realism of the expectation that is formally entertained of the undergraduate—the conventional (and official) assumption regarding the amount of reading and learning he can reasonably be asked to get done in his two or three years. It is of course only an unrealism of formal demand: it means in practice that a journalistically gifted student can refrain decidedly from overworking and get a First on a canny investment in odds and ends. (How Margaret Drabble got a First at Cambridge I won't conjecture, for though she made it public that she did *no* work, she later qualified the claim.) I myself, I suppose, might be said to expect a great deal of students. I think I do. But firstly let me emphasize this necessary condition: no one should be admitted to read English at any university who isn't of university quality and hasn't a positive bent for literary study. You may call *this* unrealistic. I hope not; for if you are not prepared to fight with unyielding conviction, and the relevant *true* realism as to standards, on this issue you may as well recognize at once that the cause is lost, and that it matters very little how we should conceive the function we are supposed to stand for. I take that as granted. I take it as granted too that the seniors in the school should be, for at any rate the most part, persons interested in literature in such a way that they take pleasure and find profit in discussing it with intelligent young students.

When I say, then, that the student should feel that he belongs to a collaborative community I mean something real by 'collaborative'. The *raison d'être* of the school won't be properly comparable to that (I invoke an influential current analogy) of an industrial plant turning out products; it will be very largely in what the community is in itself—a point I shall revert to later. Immediately it can be said, with regard to the student considered as someone undergoing Higher Education, that the problem of acquiring something coherent, meaningful and organic, a living reality that he can carry away with him (or *in* him), will obviously be affected radically, and in a very desirable way, by his collaborative membership. The 'function of criticism at the present time', to insist on an emphasis that I intend very seriously, inheres in the total collaboration, but he will have felt himself to be *of* it, with momentous consequences for the spirit of his work and his grasp of the truth that it's in the present, or not at all, that English Literature has its life.

And this brings me back to where I broke off. I was referring to the immense help that, in this matter of the relation between the present and the past and the way in which an organic (and therefore changing) English Literature exists, transcending the 'past' and 'present' of empiricist commonsense, is to be got from Eliot. He offers to deal with those themes, of course, in his best-known theoretical essays, the first two in *Selected Essays*, the first of which is widely supposed to be a classical piece of thinking ('theoretical criticism'), but they are not, where the critic is in question, what I have in mind. They seem to me pretentious, confused and unilluminating, and to exemplify a bad kind of French influence on Eliot. Their real function and use is to serve as distinguished examples of bad criticism. And this is a cue for making an important point: it is an essential aspect of Eliot's value for us that intelligent critical recognition of what he achieved can't but entail adverse and severely limiting judgments. And it will certainly be a great advantage to a student to find himself, in the opening stages of his course, distinguishing evaluatively, in a context of preoccupation that enables him to do so with conviction, between the modes of critical offer represented by, on the one hand, 'Tradition and the Individual Talent', which has been found so impressively quotable, and, on the other, the essays on the seventeenth century.

In particular, it is the one on the 'Metaphysical Poets' (a review of Grierson's *Metaphysical Lyrics and Poems*) that truly and with pregnant effect illuminates the nature, where there is a living creative continuity in a great literature, of the relation between the present and the past. I ought to add at once a reference to the help that, in another sense, is provided by J. B. Leishman. When I say that it is something for which gratitude is due I'm not just indulging in irony. That in so respectable a book as *The Monarch of Wit*, which the student will in any case read for its direct scholarly usefulness, he should be presented unwittingly by the author with a challenge to a crucial dissenting judgment is to be hailed as a piece of good luck, and we should use the challenge as a *locus classicus*. I say 'crucial' because for the student who sees the judgment as inevitable and unanswerable an insight has flashed in such a way as to become, not just a light (it is magnificently that) on the difference between real criticism and that which answers to the bad sense of 'academic', but an inner living principle for him—an active potency of perception and organizing intelligence in his work.

These are the sentences to memorize of the passage in which Leishman throws doubt on the critical value of Eliot's commentary on the Metaphysicals; they form the focal *locus classicus*:

> 'In the second place,'—Eliot having been with urbane firmness placed as unscholarly (does he not base his attitude on the good poems, whereas there are many bad?)—'both in these essays and elsewhere, Mr Eliot was writing not merely in a spirit of disinterested curiosity but (though he never explicitly admits it) with something of an axe to grind. He was writing, not merely as a critic, but as a poet, or as what he himself calls a "poetic practitioner", and a question always at the back of his mind was this: "From what earlier poets can a modern English poet most profitably learn?" His preoccupation with this largely explains both his exaltation of the metaphysicals and that denigration of Milton which he continued intermittently for the next twenty years.'

Leishman's academic blankness here matches the Philistine blankness of the Snow world. What he is dismissing is the possibility of an intelligent study of English Literature. By speaking (it is a revealing metaphor) of Eliot's informing interest as 'an axe to grind' he suggests that it's arbitrary, and inimical to critical intelligence. But in fact it is the paradigm of the sensitized percipience that any intelligent critic—and the student aims at being one—takes to the past. Eliot's 'axe to grind' was that of the poet who was qualified by gift and bent for the task to which he dedicated himself: that of proving that something *could* happen in English poetry after Swinburne.

The lines of inquiry into the nineteenth century he incites to and orientates for I will put aside—though certainly the student won't. The particular point I have to make is that the distinctive value of Eliot's criticism in these seventeenth-century essays is conditioned by its being the 'poetic practitioner's' and having the limitations of aim and scope that the description implies. They are wonderfully pregnant, but disciplined by the practitioner's (*that* practitioner's) special interest, and sharply focused: their importance for us is readily seized. It is given in two separate sentences in the essay on 'The Metaphysical Poets'. One (it comes second in the essay) gives us the directing preoccupation of the modern major innovating poet: 'The possible interests of a poet are unlimited; the more intelligent he is the more likely that he will have interests: our only condition is that he turn them into poetry, and not merely meditate on them poetically.' The other tells us what his preoccupation found peculiarly relevant to it in the seventeenth century: 'The poets of the 17th century, the successors of the dramatists of the 16th, possessed a mechanism of sensibility that could devour any kind of experience.' The phrase we need to emphasize in this sentence is 'the successors of the drama-tists': thrown out in that ostensibly casual way, it comes from the critic who is a 'poetic practitioner', and it makes quite plain what it is that Eliot valued in Donne, and registers in doing so an original critical perception that is of the first importance for us, as for every student of English Literature. It matters a great deal more where Donne is in question than such hints as 'felt thought' and the phrases in Eliot's essay that made the 'metaphysical sensibility' a fashionable theme in the days of Empson's early poems. It points us back to Shakespeare, for Shake-speare, whatever passages of the minor writers Eliot may have had in mind, was the great master in the background. And what Donne, the master of the metaphysical school, did that mattered most to Eliot the poetic innovator was to bring into non-dramatic poetry the Shakespearian use of the English language.

If Eliot owed a debt to Donne it was a debt to Shake-speare. And that he did owe such a debt becomes plain when, in the context of these considerations, we re-read that early poem of his, 'Portrait of a Lady'. It appeared in *Prufrock* during that first war, and that the obvious comment on it has not been obvious these forty years shows how little what Eliot offers has been taken. We, at any rate, can say: 'Already in 1917 this young poet had de-cisively "altered expression" ' ('expression', you recall, 'is only altered by a man of genius'). There is no suggestion of metaphysical intellectuality in the poem; nor is there anything in the manner that prompts us to say 'Shakespear-ian'. But, though the rhythms and the versification can have offered no problems to a conventionally cultivated reader even in 1917, yet Eliot's use in verse of the English language should have been seen to be momentously inno-vating. His effects, achieved with masterly precision, de-pend upon his appealing to the reader's sense of how things go naturally in the living spoken language and the speaking voice. Thus he exercises, in the only way possible, a subtle command of shifting tone, inflexion, distance and tempo. We are given occasion to reflect on the significance of his referring (in the essay on Marvell) to the preoccupa-tion of Victorian poetry with creating a dream world or poetical otherworld. The mode of 'Portrait of a Lady' en-gages the full attention of the waking mind. Moreover, it excludes from poetry no 'interest' of the practitioner's, no element of his experience. And though the poem contains no suggestion of any taxing intellectual chal-lenge, we can see in the poetic the potentiality of the strongest presence of thought: there is no paradox in the development that, starting here, culminated in *Four Quartets*.

I'm not lecturing you on Eliot; I haven't been instructing you, whatever the appearances. You see my embarrassing dilemma. I must, in my effort to put before you what, relevantly to our concern, I have in mind, inevitably work a great deal in terms of exemplary illustration. My illustrations necessarily involve much personal judgment; they are pondered and responsible, and I think them sound, but I don't forget the 'yes but . . .'. That is, I count on their receiving enough assent from you to serve my purpose, which is to define vividly (and not abstractly, which is in fact impossible) what I mean by the principle of life that should inform the student's work—and the work of the total collaboration; the principle that seeks, feels for and develops nervous structure in the developing field of study. I came nearest to putting my finger down and saying 'There you have it', when I brought together that passage of Leishman and the two crucial sentences from Eliot's 'The Metaphysical Poets'. I haven't been offering to lay down a syllabus. I know that many kinds of ordering and arranging work are compatible with the spirit that alone can justify and save university English in the world of enlightenment and Lord Robbins. I know that emphases will vary. I know too that separate provinces have to be tackled as, for the time, having the centre in themselves, and that the specialist guide is needed. Simply, the living principle, the creative and unifying principle of life, made strongly active as I'm suggesting it should and could be, will affect every patch in the total field of work and make all the difference.

I'm not, then, instructing you about Eliot. I'm pointing to the grounds for my saying that it would be absurd not to make the most of him. But I ought at once to go on to say that I also said: 'We can't do without him'. If we hadn't had Eliot we should have *had* to do without him, but it would have put us under a serious disadvantage. The truism that English Literature has its life in the present, or not at all has to be effectively asserted: it is basic. If we have to hand the means of bringing home strongly in the face of Lord Robbins, Lord Balogh and the enlightened, the force of the associated truth, that life is growth (not economic growth) and creative responsiveness to change, we should *use* them. The progressivist's acceptance of the fact of change has for corollary a contempt for tradition, conceived as a timid clinging to old habits, and implicitly posited as the only alternative. That is why I avoid the word 'tradition', and, in speaking of the need to maintain cultural continuity, insist that the maintaining, being either a strongly positive drive of life or pitifully nothing, is creative. Only in terms of literature can this truth be asserted with effect in our world, and the asserting must be, not a matter of dialectic, but itself, in a patently illustrative way, an assertion of life. And here I state the unique nature, and the central importance, of English as a university study.

The problem of ensuring that the force of the truism I have appealed to tells as it ought in the student's work —in the work of the school as a whole—points to the desirability of there being a major creative writer of our time who deserves study as such. Well, we *have* Eliot, the last—the most recent—manifestation of major creativity in our literature who died only the other day. If his supreme work, *Four Quartets*, was completed twenty-five years ago, that doesn't make him any the less, in relation to the distinctive stresses of technological civilization, a great creative writer of *our* time. We must give him in any working scheme of study the place that his claims to attention and our need indicate. For one thing, that is the only way of countering the conviction, which will certainly persist among the enlightened, that the right response at the university to change is to introduce into the English department the study of really modern literature, which means the writers who are for the time being accepted quasi-classical values in the BBC world, and the contemporary classics who are brought back by students and young lecturers from their year or two in America.

Of course, D. H. Lawrence is a much greater writer, and he is also a critic—an incomparably greater critic than Eliot. But he doesn't lend himself to our needs in the clear way in which Eliot does—doesn't, that is, in a way that helps me in my expository problem. Eliot's special eligibility for our purposes is bound up with his being so much the lesser writer. That is what I meant when I said, earlier this evening, that it is an essential aspect of Eliot's value for us that intelligent recognition of what he achieved can't but entail adverse and severely limiting judgments. So much and so clearly is that so that I must now emphasize his major quality. Major quality in a creative writer manifests itself in his being very exceptionally alive to his age and responsive to its deeper-lying spiritual stresses and sicknesses. 'Conscious' is a significant word of Eliot's —'they had not the consciousness to perceive that they felt differently and therefore must use words differently'. This sentence of his invites us to remark that, if he wrote his best criticism as a 'practitioner', a word that suggests a habit of unprofitable talk about 'technique', he wasn't as a practitioner at all like Ezra Pound. His 'alteration of expression' mattered so much because it was a product of 'consciousness'; that is, because it had behind it the pressure of 'interests' that were profound inner needs—needs of the representative kind that make an artist major. Just what they were, and how far representative, it is the business of every collaborative study of Eliot to determine, and in the pursuit of the business students will be inquiring into the profounder problems of our civilization—the problems that sociologists, social scientists, social workers, anti-racialists, statesmen, and the enlightened in general ignore.

I will merely, in order to further the definition of my theme, make minimally two or three points. The profound response to our civilization that makes Eliot a major creative writer is misrepresented by those who perpetuate him as the poet of *The Waste Land*. The corrective is to point to 'Difficulties of a Statesman' and then to Section III of 'East Coker'; together they give the cue for saying what it is that, central to the Eliotic preoccupation, is at the sick deep centre of the modern psyche and may fairly be called the technologico-Benthamite plight:

Cry cry what shall I cry?
The first thing to do is to form the committees:
The consultative councils, the standing committees,
    select committees and sub-committees.
One secretary will do for several committees.
What shall I cry?

Cry?—'participation', obviously. I haven't time to do what I had in mind to do—read the 'East Coker' passage ('O dark dark dark') that gives the personal-impersonal depth of involvement and the full pregnancy. What we have there is what troubles us all, whether we know it or not —except, perhaps, the computer-addicts and the Annans. No, even them. It's the loss of ends and significance in the complication of the machinery: all ends having authority lost, all but those lending themselves to statistical mensuration—as reductive 'equality' *does*. And to read Eliot's poetry, his major poetry, is to be made to realize in what sense a serious pondering of the plight must necessarily be a matter of thought and exploration at the religious level. And this is the moment for reiteration: above all when confronting the major Eliot does one find oneself compelled to the 'yes, but—'.

Here we have an essential condition of his appropriateness for our purpose: he prompts, powers and illuminates fundamental exploration, but never puts the real reader of his poetry in danger of mere acceptance. Judgment will be tentative, and there will not be a unanimity of tentative judgment in any summing-up group. Conveniently small in quantity as the *œuvre* is, the Eliotic phenomenon is challengingly complex. My own relevant testimony is that, in the last two or three years I have written three separate critiques of his poetic achievement, different, but not contradictory, and I feel that I might write two or three others. That, for me, is the only way of dealing with him. Undoubtedly, any group paying close attention to his work would be very conscious of the contrasting Lawrence as a background, and the consciousness wouldn't be critically inert or unexpressed. I don't think that the curricular economy I myself should sketch would find room for any set formal study of Lawrence. Nevertheless, I could suggest manageable ways of bringing the fundamental contrast to bear on the placing of Eliot, his limitations, his validities and his invalidities.

But there is still something essential to note regarding the seventeenth century and the characteristic immense service Eliot has done us there: I've said nothing about the 'dissociation of sensibility'. That phrase is now uttered, or written, with an effect of irony, the implication being that we know it now as thoroughly discredited, an unscholarly vagary of the poet that was seized on by the more unscholarly and less excusable advanced critic—which shows, I think, how essential it is that University English should make the most of the insight the phrase registers, the incontestableness of which has been hidden from the professionals by its truth and its pregnancy together. Incontestableness—for does anyone question that in the seventeenth century a momentous total change in civilization took place, so that by its close we have the modern world, launched on its accelerating advance towards the consummation we now know? Eliot's phrase, and I can't think of any that would have done the job better, calls attention to the change as it is manifested in the English language—and manifested so notably in the literary use. It is what hits us so disturbingly when we see what, with sincere deference, Dryden did to Shakespeare in *All For Love*. Could any profounder or more essential change take place, and so rapidly, in a civilization?

The change in literature gives us a magnificent focal line, or (you can say) control, for a study—and the literary student is as such committed to what of that kind may be possible—a study of the total change in its various aspects. If that responsibility is taken seriously, as it must (in relation to our vital modern function) be, some relevant specialists stand to profit by the *liaison* contacts that should be sought, for there is no historian, or philosopher, or political specialist whose thinking wouldn't benefit if he could make an initiatory acquaintance with the kinds of thinking that belong to literary study.

I have raised the supremely important head of *liaison*. Its importance is not to be measured by the amount of time I give to it. In fact, the quantitative criterion has no authority in the realm of our essential concern, life—life itself, which, not being able to engage on it, the invoker of the criterion always leaves out as not really real, or too important to matter, or too axiomatic to count. What we stand for, being (to put it negatively) non-specialist, is, positively, something of which in a university milieu we can with proper optimism say that everyone stands humanly, and that is vitally, in need of it. Some specialists essentially need as such to be fully human, and can be part-disabled by the life-deficiency that impairs vision. I'm not preaching wild optimism. I don't for a moment suppose that you can leaven the whole quantitative lump of a university. But where there is the vital centre I'm envisaging, and the fostered transcending of departmental boundaries, you will have beyond the boundaries some understanding of why the centre is vital. Shading outwards beyond that there will be some awareness that humanity can't safely ignore *humanitas*, and further out still, shading off into not unkindly near-indifference, some tacit and vague respect. What I am describing—and don't take it for a Utopian dream—is a university that tells in the life of the country as itself a centre of the informed and responsible opinion that an educated public would make into a climate in which politicians, bureaucrats, Vice-Chancellors' committees and Ministers of Education had to do their planning, negotiating and performing.

I obviously haven't time to convey my sense of the practicality of *liaison* by illustrating the diversity of forms it can take. I will confine myself to an indisputably crucial field where, if anywhere, the opportunities for it must be seized and cultivated—crucial, because a proper aliveness to the fundamental nature of the world's present sickness depends upon some knowledge and perception in relation to that field, and these entail an understanding of what happened there in the seventeenth century when the great change became irreversible. I am thinking of course of the confident start of science upon its accelerating advance. What the student needs to acquire a minimum knowledge of is the way in which the 'common-sense notion of the universe' (Whitehead's phrase) took possession of the ordinary man's mind, and with what consequences for the climate of the West and the ethos of our civilization. This involves being able to state intelligently what the Cartesian-Newtonian presuppositions were and to what kind of philosophical *impasse* they led—that still exemplified in the philosophies of science and the positivist and empiricist fashions that prevail.

Clearly it should be possible to enlist help from the Philosophy department. But the English School itself shouldn't be helpless to detect an empiricist bent, with the attendant limitation of his wisdom and use, in the available specialist. That means that there should be at least one person in the school ready to read and discuss with students (say) Marjorie Grene's *The Knower and the Known*, or key parts of it. Where the conditions were ideal there might very well be a small group of students being helped by a qualified guide to appreciate the significance of Whitehead, Collingwood and Michael Polanyi. Such a group might be itself a *liaison* group (I mean, at student level), bringing profit to both elements.

I must say no more about *liaison*: I can only hope I have said enough to suggest its importance. One can't foresee where an interest that an *ad hoc* contact may start will end.

It's very possible that the cooperative philosopher will discover interests that have a vital relevance to his thinking in his own field. For the literary student, in any case, there will be a close relation between the inquiries I've just touched on and a grasp of the significance for us of William Blake. Blake is the protest of life against the world of Newton and Locke, and it is in relation to him that the student takes the force of the shift that, in this linkage, has substituted Locke for Descartes. In Locke's epistemology 'ideas' have become 'impressions', and 'impressions' (if visual) are images registered on the passively recipient retina. This, as Blake insists with all the force of the creative affirmation, is to deny the essential creativeness of life, and to be committed, therefore, to repressing life itself. Perception, he insists in art and aphorism, is creative, and there is a continuity from the creativeness of perception to the creativeness of the artist.

You must forgive my appearing to be instructing you about things you are not ignorant of: I don't know how otherwise to make the absolutely essential points. My point at the moment is that, whatever ordering of work you adopt, you must recognize on your chart of the underlying organic structure—which, of course, won't steadily remain merely tacit—the key importance, for any English course that is duly informed with a sense of the present, of Blake—who points forward to the Dickens of *Hard Times* (and not merely of *Hard Times*) and the Lawrence who, with Russell's blank egalitarian enlightenment to fight, invented the word 'disquality'.

To emphasize creativity as Blake did is to be committed to bringing home to the world, if in a world of Lockean or technologico-Benthamite blankness that can be done, that you can't generalize life, that individual lives can't be aggregated or averaged, and that only in individual lives is life 'there'. The fact that Locke was, in acceptance and actuality, the philosopher of commonsense made him the enemy in a way that ensured Blake's constantly naming him as such—coupled with Newton. For he answered perfectly to the basic canon of the civilization that was to have its assumptions registered in *The Tatler* and *The Spectator*: the canon to the effect that nothing intellectual made sense, and nothing really mattered, that wasn't within the compass of the ordinary civilized person as such. But, for my particular approach of the moment, I should have said, 'To the Gentleman *qua* Gentleman'. The culture was not only commonsense, but 'polite'. That is, its conventions of expression and all its aspects were in close touch with a code of elegant manners.

What I am calling attention to is the unprecedented insulation established at the end of the seventeenth century (an aspect of the total change) between the 'educated' and the popular. It is implicit in the change Eliot points to with his phrase, 'dissociation of sensibility', but he doesn't seem to notice it as such. You will recall the way in which he refers to the metaphysicals as the 'successors of the dramatists', but doesn't mention Shakespeare, the great relevant fact—source of life—in the background. Nevertheless he does implicitly, in relation to an intensely realized present, refer us back (and can it have been quite unconsciously?) to Shakespeare. And when we consider his creative achievement and the light thrown on it by his criticism, we find that he takes us back to Shakespeare in a way that has a peculiar relevance to the problem we are faced with in having to define and enforce the function of the university in the technological world. That he should do so partly in spite of himself—by making us realize in

full consciousness what he himself doesn't—is characteristic of his full value to us.

His limitation comes out in the third paragraph of 'East
Coker', where he makes it plain that for him the country-
folk of pre-industrial England are yokels. In the interests
of economy I'll read a few sentences from my Cheltenham
address, where I took pains to make the necessary points
with extreme brevity.

'Yet the country-folk whom Eliot reduces to this created
the English language that made Shakespeare possible. . . .
It developed as the articulate utterance of a total organic
culture, one that comprehended craft-skills of many kinds,
arts of living formed in response to practical exigencies
and material necessity through generations of settled
habitation, knowledge of life that transcended the experience of any one life-span, subtly responsive awareness of
the natural environment. Of course, it had always been
affected by higher cultural influences of the kind that
commend themselves to Eliot's recognition. But the native
language—it was the language of the native priesthood—
wasn't weakened by such influences, but strengthened; it
responded livingly, and didn't become less English. By
the time Shakespeare was discovering his genius there was
ready to his hand a vernacular that was marvellously recep-
tive, adventurous and flexible, yet robustly itself; capable
that is, of accommodating and making its own all the in-
fluxes of the Renaissance.'

Eliot, after all, was a fellow-countryman of Pound: he
couldn't, in his imaginative thinking, conceive—in spite
of the great fact of Shakespeare—of a sophisticated art
that should grow out of a total organic culture.

'By the close of the 17th century the conditions of
Shakespeare's kind of greatness had vanished for good.

Shakespeare could be at one and the same time the supreme
Renaissance poet and draw as no one since has ever done
on the resources of human experience, the diverse continuities, behind and implicit in a rich and robustly creative vernacular. By 1700 a transformation as momentous
as any associated with the development of modern civilization had taken place, never to be reversed. The new
Augustan culture represented by Pope and *The Tatler* entailed an unprecedented insulation of the "polite" from
the popular. There *could* be no reversal: the industrial
revolution, which by the end of the 18th century was well
advanced, worked and went on working inevitable destruction upon the inherited civilization of the people. Dickens
was the last great writer to enjoy something of the
Shakespearian advantage.

'What has been achieved in our time is the complete
destruction of that general diffused creativity which main-
tains the life and continuity of a culture. For the industrial
masses their work has no human meaning in itself and
offers no satisfying interest; they save their living for their
leisure, of which they have very much more than their
predecessors of the Dickensian world had, but don't
know how to use it, except inertly—before the telly, in
the car, in the bingo-hall, filling Pools forms, spending
money, eating fish and chips in Spain. The civilization
that has disinherited them culturally and incapacitated
them humanly does nothing to give significance . . . to
their lives, or to any lives. Significance is a profound
human need, like creativity, its associate. The thwarting
of the need, or hunger, has consequences not the less
catastrophic because of the general blankness in face of
the cause. The complacent "understanding" with which
the enlightened contemplate these things is not under-

standing or enlightenment, but merely a manifestation of the disease from which our civilization is suffering. When the new maturity claimed by Youth Club leaders for today's young has established its right to act out an intuitive new wisdom independent of any derived from past human experience, the achievement of happiness won't have been advanced, however confident the expectation.'

There is no going back: no one will suppose me to be implying otherwise. But we have to understand our plight in order to realize what kind of effort we have to make. There is no restoring the old kind of general continuity-maintaining creativity. But it is fatal to let the cultural inertness of the technological age spread and prevail till anything else is forgotten and incredible. It already nearly is (witness T. S. Eliot), and we have for brutal manifestation statesmen who regard a university as an industrial plant, students who can feel that their claims to prescribe their curricula and examine themselves are regarded by smiling Vice-Chancellors as discussable (the students themselves being products of pop-art and the culture of the telly), and Provosts of King's College, Cambridge, like Lord Annan and Dr Leach. What we are rapidly heading for is the hopelessness of America, which now *knows* in its deep psyche that it won't be happy even when it has solved its racial problem, achieved a New Deal, landed men on the moon, and made a good start towards beating Russia to the next planet.

What we have to look to, what we have to ensure and power, is the maintenance of cultural continuity by a body of the educated—of those who are conscious of the general need. The nature of the effort by which we must do that has been my theme. My 'must' doesn't portend optimism or any form of simple-mindedness. I'm not supposing that

English Schools such as I have described are likely to be realized in all, or half, or even a quarter of the universities. Nor do I suppose that even one university will be, as an unqualified totality, the centre of consciousness and responsibility I've pointed to as the paradigm. But if even one university became at all known as a focus of creative life and *humanitas* in the way I've been trying to suggest, that would be immensely worth achieving. That university would, to its great advantage, have a reputation. And it would be a source (and a recipient) of strength and encouragement in relation to less successful efforts of life in others. If there were a number of universities, each, inevitably, with its own spreading network of relations, then there would be the educated public without which there is no hope.

I've made it plain, I think, that 'educated public' doesn't say all. The universities, the creative nuclei, must create something that the phrase doesn't by itself suggest, and without which the sickness that figures as student unrest, violence, drugs, sex-addiction, absenteeism and Lord Annan won't start to be cured. If one has to answer the question, 'What is the something?', 'creativity' is about all it would be wise to say—though perhaps one might add that creativity creates significance and reality. Certainly, if the kind of effort I've been talking about got going at all strongly, so that the country could be said to believe in it, then the state commonly deplored as 'lack of a sense of purpose' or 'loss of a sense of national identity' would be on its way to being remedied.

Let me say at once, to separate myself finally from all suggestion of enlightenment, that I don't like the attitude to our imperial past that makes it merely something to be atoned for—I say that as the scion of a line of Little Englanders. On the other hand I have no sympathy with

those who hanker after lost 'greatness'. That kind of greatness belongs to a past era, and has gone for ever out of
the world. That is what America has discovered—portentous development!—in the past few years: the acquisition of unrivalled power hasn't brought, and can't in
our world bring, greatness.

And I come now, as seems to me necessary before I
close, to define my attitude, which I think the right one,
towards America. I think it misleading to describe me as
anti-American. True, I'm prepared to say that we should
do everything we can to save this country from Americanization: it's horrible to think of it as being, what it is
rapidly becoming, just a province of the American world.
But I know Americans who agree with me, and I think
that they and their like (of whom there are after all a good
many) are the truest friends of America. I don't confidently
expect them to applaud with unanimity when I say that
hope of salvation for America depends upon our success
in the creative battle here, where we can still open it, and
wage it, and resolve to win (or not to lose)—but some of
them might. If this sounds patronizing and unrealistic,
let me say that I am not indifferent to the question of
power, any more than I am hostile to the existence of
politicians and governments. And that America is *not* less
strong than Russia seems to me distinctly a matter for
self-congratulation on our part. When I ask what hope
there is for humanity in Russia, I turn cold. If this country
could generate in a decisive way the kind of creative effort
I have described as especially *our* business, that would be
its true greatness in history reaffirmed—perpetuated by
renewal in terms of the modern world, and nowhere
would it receive more heartening recognition than in
America. Of that we can be sure.

V

'Literarism' versus 'Scientism': the Misconception and the Menace

'LITERARISM' VERSUS 'SCIENTISM': THE MISCONCEPTION AND THE MENACE

'THE computer can in no way lift the responsibility from human shoulders.' That reassuring statement caught my eye on the front page of the first issue of the *Times Literary Supplement* for this year (January 1, 1970): there it was, not far from the beginning of the article. 'Responsibility' is an important word. Unless we alter its meaning altogether—and such a change, if fully accomplished, would be an important event in the history of civilization—responsibility must, of its nature, *be* human; but the presence of the adjective in the sentence I've quoted, where it's not in any case superfluous, is essential to the reassuringness. Actually, however, I wasn't altogether reassured. Such reassurances are characteristic of the phase of civilization in which we live—they may be said to be *de rigueur* in commentaries on coming advances in civilization addressed to the enlightened. I remember, for instance, reading in the *Guardian* a leader that concluded with the remark that, of course, it wasn't enough to think of the reason for the accelerated multiplication of universities as being the country's need, in this technological age, of rapid progress in science and scientific education: 'Science is a means to an end.'

'Science is a means to an end': that emphasis might have been accepted as a hopeful sign if one had, in the *Guardian*, ever come across any evidence that in the *Guardian* milieu the 'end' in view was ever seriously conceived as anything other, anything more adequate, anything less congenial to statistical criteria, than a 'rising standard of living'—an advancing G.N.P., more equitable distribution, and an improved Health Service. Please don't suppose that, in this matter, I mean to discriminate against the *Guardian*. An essentially similar report to that I have intimated would have to be made on the *New Statesman*, the *Spectator*, *The Times*. That's where we are; there you have the moment of history at which we live.

The idea that some reassurance might be in place as to ends hasn't been forgotten; the reassurance, accordingly, is thrown out—or thrown in; but the idea of its being required that it should mean something, seriously conceived and really meant, *has* been forgotten. Thus the Prime Minister, making a major pre-election speech, holds forth on the great constructive achievements of the Government, the greater achievements to come, the planned increase of wealth, the re-structuring of society, the determined democratic drive in the educational field—especially in higher education, and comes duly to the point at which he says, with all the grave impressiveness at his command: 'But we must not forget *quality*.'

'Quality' is a word. If the word, in such a context, is to have any substance of meaning, it must be a meaning that entails a reference—one that is intelligent, convinced, and can be taken seriously—to human ends. Actually, the word is left to do its work as a mere word, unsupported; its work is to stand up against, to counter reassuringly, the clear implication of the context. The implication is that we need take no ends into account in our planning and calculating but those which are looked after by the quantitative criteria, the statistical: 'quality', that is, will look after itself. Clear implication? 'Clear' isn't, perhaps, the right word; it might suggest that in any educated company I could expect general concurrence in my attitude.

But I'm not at all meaning to make a point against Mr Harold Wilson in especial. You'll find the same essential indifference to what is, after all, a very important order of consideration in any statesman or politician of any party. I don't myself confidently believe that when an enlightened and cultivated Conservative statesman leaves the political field for the academico-political, *his* voice will be steadily heard, loud and clear, saying unequivocally the necessary challenging unanswerable things in the ominous days of battle to come—necessary, if the battle is not to be abjectly lost, while the the world remains blank to what is at stake; the battle that is *likely* to be lost, the defeat being hailed as a victory for enlightenment.

My point, one implicit in my title, is that the educated and cultivated have, in general, given in—have surrendered to the climate of the technological age. I say 'implicit in my title', because I don't offer the formula, 'Literarism *versus* Scientism', as my own. The term 'literarism' was in fact coined by the late Aldous Huxley for use against me, and I quote it as representatively symptomatic—as signalizing a characteristic confused vagueness of thought in a well-known intellectual, the vagueness expressing an intention that really portends surrender. I don't think that Huxley was in any way a distinguished mind, but he was cultivated, widely read and voraciously inquiring, and he enjoyed a general reputation as superlatively intelligent. For all I know, he had met Snow and was on friendly terms with him. In any case, he chose to balance Snow's 'scientism' against a 'literarism' attributed to me: we were equally deviations from the centre of truth. What, positively, *was* at the centre of truth Huxley didn't, I thought, make at all plain, except that *he* was there.

Now, there's no problem about the meaning of 'scientism'; it's Snow. His insistence on the supreme importance of science in education (for his position amounts to that) goes with a dismissal of the great creative writers of the nineteenth century and later as 'natural Luddites'— whether he knows it or not, that description, with its context of assumptions, *is* a dismissal. There's perhaps no reason why we shouldn't read them; they have, one gathers, what is claimed pre-eminently for Dickens; entertainment value. But we must recognize unequivocally that the interest they offer has only the most dubious bearing on mature thought, serious thought, about the problems facing mankind in the technological age. They are natural Luddites; creative writers—those exalted by literary persons—are like that. The truly significant novelists are significant in the way in which Zola, H. G. Wells and C. P. Snow are. They have their ideas, based on a real understanding of the mechanisms of civilization, and their fictions enforce, put over, their ideas.

That's the attitude; that's what Huxley points to as 'scientism'. Well, science is obviously of great importance to mankind; it's of great cultural importance. But to say that is to make a value-judgment—a human judgment of value. The criteria of judgments of value and importance are determined by a sense of human nature and human need, and can't be arrived at by science itself; they aren't, and can't be, a product of scientific method, or anything like it. They are an expression of human responsibility. In the rapidly changing external civilization of the technological age it is peculiarly necessary that the consciousness of human responsibility and what it involves should be cultivated and strengthened to the utmost—that there should be a directing sense of human need and human ends the most richly charged with human experience that can be made to prevail. To suggest otherwise is to propose leaving the human consequences of the process of change (which more and more manifestly have their own momentum and internal logic—which is often *anti*-human) to be determined, in so far as human motives are involved and tell, by the crude promptings of a starved and perverted sense of human nature—the sense we see being generated around us.

All this is obvious enough, you'd think; at any rate, you have there my position—in which Huxley, with his imputation to me of 'literarism', didn't concur. Just what third, what truly central, position he defined as his own I haven't made out; but I haven't tried very hard. What he seems to me to exemplify is the failure of the cultivated —more generally, the non-specialist educated—to see the issues clear and to be the allies they surely ought to be. As far as public demonstration went, as a matter of fact, I found them, on the occasion of my ill-famed Richmond lecture, enemies: week after week following the appearance of the lecture in the *Spectator* both *its* corresponding pages and the *New Statesman's* were charged and swollen with letters—and (I suspected) telephone communications— from the intelligentsia, denouncing me and my cruel, gratuitous and stupid assault on poor enlightened Snow, who, thus reassured, took cover under magnanimity. I got, of course, private letters of support, too; a good many from scientists, British and American, the refrain of which was: 'Thank God, someone has said it at last!'

Nevertheless, what I have to report is failure. The failure is given you in the established habit (they impart it in the

Scholarship Sixth) of referring to the 'debate about (or between) the Two Cultures'. There has been no debate. And what I pointed out was that there's only one culture, and that Snow merely—and symptomatically—abuses the word (a very important one) when he talks about a scientific culture, generated out of the technical knowledge and the specialized intellectual habits that scientists have.

It's the general blankness in face of the issues that's so discouraging. I was, I confess, a little amused when, sitting at a formal lunch next to the director of a City Art Gallery, I was told by him, in the tone of one saying something very impressive: 'A computer can write a poem.' I replied, very naturally, that I couldn't accept that, adding that it was one of the things that I *knew* to be impossible. When he responded by being angry, fierce and authoritative, I reflected that he was German, if an émigré, and that in any case his business was *Kunst* and he hadn't said that a computer could paint a work of art. The other occasion on which I was confronted, point-blank, with the preposterous and ominous claim, which by then I had discovered to be pretty current, it made a profound impression on me. The testifier was a philosopher, a lady and cultivated; her place and conditions of residence gave her access to a friendly computer laboratory. She had taken advantage of the opportunity, I gathered, to develop an intense experimental interest: 'It's incredible,' she said, 'what a computer can do; it's awfully fascinating; you know, a computer can write a poem.' I couldn't let that pass; with the appropriate urbanity I said: 'Well, "poem" means different things.' There was no Teutonic anger this time. There was a sudden descent, a heightened nuance of pink, a concessive philosophic laugh, and then; 'O well, yes; but it's great fun.'

That any cultivated person should *want* to believe that
a computer can write a poem!—the significance of the
episode, it seemed to me, lay there; for the intention had
been naïve and unqualified. It *could* be that because of the
confusion between different forces of the word 'poem'.
And yet the difference is an essential one; the computerial
force of 'poem' eliminates the essentially human—eliminates human creativity. My philosopher's assertion, that is,
taken seriously, is reductive; it denies that a poem is
possible—without actually saying, or recognizing, that.
If the word 'poem' can be used plausibly in this way—
and by 'plausibly' I mean so as to be accepted as doing
respectable work—so equally can a good many other of
the most important words, the basically human words.
Asked how a trained philosophic mind in a cultivated
person could lend itself to such irresponsibility, I can only
reply that the world we live in, the climate, makes it very
possible.

It is in the light of these reflections that one reads the
article on the computer in the *Times Literary Supplement*
(and note that the *Times Literary Supplement* is a literary
journal, addressing persons of non-specialist cultivation).
The article contains some propositions,, some general
constatations, some statements of principle that in themselves invite concurrence as wise, sound and obvious.
There's the one I started by quoting: 'The computer can
in no way lift the responsibility from human shoulders.'
Responsibility—yes, a very important word; and we
watch to see just how the writer conceives responsibility,
and in what ways, in his vision of a prosperous human
future, the conception would be realized. That the conception needs to be both strongly positive and subtle, the
means and modes of its realization correspondingly potent,

his account of the critical human situation today and the menacing future makes plain:

> We shall have to reconsider the part played in our lives by leisure, for we shall be surrounded by it; and with education, work and leisure totally restructured the whole pattern of society will have to be changed,—for the better, if we have used forethought, almost certainly for the worse if we have not, or if the thought has been wrong.

Well, there we have an admonition that is unquestionably well-founded. Most certainly, immense changes are taking place; most certainly they will go on and accelerate; and, if there isn't forethought, if there isn't a sustained and organically collaborative effort of life-charged and life-serving human intelligence, most (not almost) certainly the changes will be for the worse. But as one looks through the article it becomes plain that the writer, when he uses the word 'thought', is no more thinking really of thought than my philosopher was thinking of a poem— that is, really meant *poem*—when she said: 'A computer can write a poem.' What in fact we have in the *Times Literary Supplement* article is a larger illustration of the cultural phenomenon represented by that episode. The essayist wouldn't have *said*, 'A computer can write a poem', or thought of saying it, because his insistence is on the computer as a servant of human life. But, without knowing it (and that's the frightening thing), he empties the reality out of life, just as the philosopher emptied the postulated thing portended by 'poem'. They both eliminate human creativity. And there is no more thought in the *Times Literary Supplement* article than there is poetry in the philosopher's poem. The writer, in the way of which Snow's Rede lecture may serve as a classic, is carried forward on the swell of cliché pseudo-thought that for journalist-intellectuals *is* thought, and doesn't notice the emptiness of meaning, or the actual significance.

The word 're-structured' in the passage I last quoted—'with education, work and leisure totally re-structured the whole pattern of society will have to be changed'—is ominous. Already life tends to be thought of in terms that make 're-structured' conceptually congenial in relation to it when essential change is in question—for leisure *is* life; it's that, or nothing. Society is overtly, and almost universally, thought of merely as something susceptible of re-structuring. Of course, I myself don't dispute that there *has* to be the approach that defines its problems, and deals with them, in terms of statistical data, charts and the computerizable generally; we are committed irrevocably to the necessity of government, government departments, complex machinery of administration, and bureaucrats. But I believe also that the 'social' has to be conceived in another way, and that nothing is more urgent than to insist on that (not at all an easy matter in the world as it is); and that we have to fight resolutely, fiercely and intelligently for the essential conception, and to ensure that it shall be at least in a minimally sufficient measure realized.

Pondering this concern—the concern for a conception of society, life and *humanitas* that doesn't eliminate the depth in time and the organic, one thinks of education as an immediately relevant theme. The business of education, of course, can't be covered with a simple formula; but, we know, there is one aspect that, in our time, it will hardly be unnecessary to insist on. It wasn't to be expected that, even in the *Times Literary Supplement*, an essay on adjustment to the computerized world would give anything like satisfying attention to that aspect, or function. The actual performance ignores it—ignores it so insistently and so completely as to provide a strikingly significant illustration of the world we live in; the world in which a cultivated philosopher can take pleasure in assuring one that a computer can write a poem. I say 'ignores insistently' because the world 'education'—which occurs, along with 'work' and 'leisure' in the passage I've quoted—appears in the essay very frequently; so frequently that education, for anyone casting an eye over the text, seems to be the writer's main interest. The significance lies in the unconsciousness of the insistence. Actually, he shows no interest in education at all; though he would probably be surprised to be told so, he says nothing about it. But in his unconsciousness there is implicit the conception of education—and it covers the university—that is being propaganded militantly by Mr Fowler, a former Minister overseeing higher education, Mr Harold Wilson (in so far as he has attention to spare), Mr Christopher Price, Lord Annan, ex-Provost of King's College, Cambridge, and actual Provost of University College, London, and the authoritative planners who, confident of general support from the electorate, plan to double the number of university students by 1980 (and at comparatively little cost).

The conception—it hardly amounts to that but there's nothing else—is what is given here: 'Computerized teaching systems will make available the world's finest teaching to any child within reach of a communications system.' This *is* education; at any rate, it doesn't occur to the essayist to go beyond this. The statement, of course (there are a number like it), has a context that extends the explicitness a little, and the context is what we have here (this statement, too, being driven home by a number on the same pattern): 'These computing systems will form

an interlocking network of information retrieval and processing systems well able to master the information explosion and the demands of any educational set-up.' It
will bring out the force and intention of the assumptions
so confidently in possession here if I read out the two
sentences that follow:

> With this network established man will have passed from the
> industrial age into the cybernetic age, and will have to re-think
> his approach to education, employment, leisure and society at
> large. He will have to re-think his approach to education
> because the computer will gradually control all structured
> tasks, whether they be the production of goods or the carrying
> out of commercial procedures.

The approach, it is plain, will impose, universalize and
rigidify the implicit notion, crude and brutally uneducated
as it is, of the reality: there will be *no* re-thinking, no
thinking at all, and the possibility of the kind of thought
made so desperately necessary (even on the essayist's own
account of the problem, though he doesn't know what
the most important phrases in it really mean) will be
eliminated—eliminated as a conceivable influence on development. What 'structured tasks', for instance, are involved—could be, or should be—in the study of English
literature? As for teaching, the word in the sense it has in
the computerial connexion applies so little that I myself
don't like using it in referring to my relations with younger
students. 'English', therefore, the discipline, the humane
addiction, the school of intelligence that we, who know
our grounds, are committed to defending and promoting
as an essential university study, doesn't, for the given conception of education, exist.

It isn't, of course, merely 'English' that must recoil
from a conception so lethally uncongenial. And I don't

suppose that those concerned for the sciences in the universities will accept it as adequate, or accept what they can't but know to be its actual politico-practical implications—for they are familiar with it, whether the *Times Literary Supplement* has come their way or not.

For the essayist is decidedly of his time, and his conception, for him axiomatic and in the nature of things (as his matter-of-course way of associating educational problems with those of 'the production of goods and the carrying out of commercial procedures' beautifully illustrates), is that which expresses itself in the more and more matter-of-course view that a university is so much plant that should be kept in full production all the year round, its staff made to *earn* their salaries, and its management governed by strict cost-efficiency considerations. This is the conception implicit in the proclaimed official policy of doubling the number of university students by 1980. Cost-efficiency: the plant is there: let us apply the criteria and the methods with simple-minded resolution and the privileged obstructiveness of the academic mind (for resistance reduces to that) will be overcome with ease.

In say, Mr Christopher Price, this, one may hazard, is the brute ignorance that might equally be called innocence—the innocence of 'students' intent on *being* 'students'. But Mr Fowler, Minister of State overseeing higher education—I wouldn't pronounce about *him*. I know nothing about him, but when one comes to politicians in that measure successful one has to reflect that the distinction between innocent and not innocent may be no simple matter. Was my philosopher innocent or not innocent in her assertion that a computer can write a poem? Mr Fowler told a conference at York recently: 'If student numbers should double by 1980, Britain would have moved a considerable distance towards the sort of mass education system that exists in the United States.' He added: 'There is nothing to suggest that we should stop at that point.' He was explicitly talking about universities—*the* universities. And there is everything to suggest that he has Lord Robbins of the Robbins Report with him—Lord Robbins who in an address (which he published in a book of them) went out of his way to remark: 'We all know the kind of person who looks down his nose when you mention American universities'—this being aimed at critics of Lord Robbins and the ethos he represents. I recalled, when I read it, the American acquaintance who remarked to me how horrifying it was to an American to see us in this country plunging into the courses from the consequences of which America was struggling desperately to escape.

Lord Robbins is a tough-minded re-structurer, and Mr Harold Wilson, whose sympathy Mr Fowler can clearly rely on, is a pre-eminently successful politician who—essential condition of success in politics—has developed no perception for what is being destroyed. But what are we to say of Lord Annan, who edited a selection of Matthew Arnold's prose for the 'World's Classics', and is the indefatigable ally of the destroyers? Destroyers?—yes, undeniably, and often outspokenly, the destroyers of standards; contemptuously and righteously very often—like Mr Christopher Price, who contends that the 'comprehensive' principle should be extended to universities, and dismisses with no argument but scorn the idea that, when democracy (or anti-divisiveness) in the accepted sense prevails, there will—or in the nature of democracy *could*—be students facing university requirements whom it would have been desirable to exclude as not qualified for university work. Are we to say that Lord Annan also is naïvely righteous when, in his public utterances, he endorses with the cultural prestige of an ex-Provost of King's the idea of a university as an industrial plant that should be kept running to capacity all the year round, and warns the actual universities that they will incur attack as citadels of privilege if they don't 'lay on residential courses' in the summer to cram children in order that they may pass their qualifying examinations in spite of not having been prevented from leaving school at fifteen?

It can at any rate be said that he enforces his injunction with menacing righteousness—'menacing' is the word. Lord Annan, in fact, as combined academic power and aspiring politician, brings us to a more obviously menacing form of the phenomenon represented by my computerologizing philosopher. I came on another instance when, opening, not long ago, a number of the New York intellectual review *Commentary*, I found my eye caught by a sentence in which Mr Raymond Williams informs the American public that F. R. Leavis is opposed to the extension of higher education.

And—the important consideration—it's essential to an adequate sense of the disease and the problem that one should recognize the way in which a climate of accepted enlightenment can make indignant righteousness out of irresponsibility, and a flank-rubbing consensus enable people to distort the plain truth with a licensed—and that's as good as moral and self-approving—recklessness. *I'm in favour of extending higher education to the utmost.* You have there sufficient reason for my insistence, which should at some time have penetrated to Mr Raymond Williams's focus of recognition, that it's disastrous to identify higher education with the university. It's disastrous because the more you extend higher education—and especially in an age of technological aids and Open Universities, and of innocents like Mr Christopher Price, and non-innocents like Lord Annan, and of ministerial authorities charged with promoting culture—the more insidious becomes the menace to standards and the more potent and unashamed the animus *against* them. Unless standards are maintained somewhere the whole community is let down, and higher education itself is not exempt from the consequences. The only place where standards *can* be maintained is the university properly conceived—the university as Lord Robbins, Lord Annan, Mr Fowler, Minister of State overseeing higher education, and Mr Harold Wilson are committed to destroying it (and let me add at once that I see no grounds for counting on anything positively better from Mr Heath's party or Mr Jeremy Thorpe's).

Scientists too will have to fight, but in *their* defence of standards they may count on the essential measure of success, for science is recognized as nationally important. And standards as we in 'English' are concerned for them are of their nature not amenable to effective presentation and assertion. They are patently not susceptible of reduction to quantitative, mathematical or any kind of demonstrative terms, and the drive of our triumphant technologico-Benthamite world is not merely indifferent, but hostile, to the human creativity they represent.

It is when we come to considering their nature—what they are, and how, if at all, they can be in our time an effective presence—that we have to make an essential point about the idea of a university. A university can't be adequately conceived in terms of what the word 'education' ordinarily suggests; a real university is a creative centre of civilization.

The prompting to the line of thought that leads here might have been found in the subtitle to that article in the *Times Literary Supplement*; it runs: 'Learning how to make the most sensible use of the computer.' 'Use', there, is the important, the challenging word. My criticism of of the article isn't hostility towards the computer; it is that the author, without being aware of it, presents us with a complete and frightening absence of the thought that the word should have portended—frightening, because the article is so representative, and appears in the *Times Literary Supplement*. I chose my own title, *'Literarism' versus 'Scientism': the Misconception and the Menace*, because it gave me the opportunity to make plain that I neither believe in any special 'literary' values nor am hostile to science. I am not, as—in a public lecture now extant in print—I tried (without success) to make it impossible for the voices of *New Statesman* enlightenment to present me as being, a Luddite. As, I think, is fairly widely known, D. H. Lawrence seems to me a voice of wisdom, human insight, and sanity, and I will quote once more the passage of him I quoted then:

> But why so much: why repeat so often the mechanical move-ment? Let me not have so much of this work to do, let me not be consumed overmuch in my own self-preservation, let me not be imprisoned in this proven, finite existence all my days.
> This has been the cry of humanity since the world began. This is the glamour of kings, the glamour of men who had the opportunity to be . . .
> Wherefore I do honour to the machine and to its in-ventors.

This was a long half-century ago, since when our civili-zation has advanced along its path (which Lawrence con-templated with disquiet) at an acceleration—and he, an intensely vital artist, was potently and inveterately creative.

By 'opportunity to be' he meant opportunity to live. And he knew that living, in the individual where only it *can* be, is an art, and one that is not *merely* individual, but depends on a heritage of arts of living that is kept alive and responsive to change in collaboratively creative use as a language is. That is the significance of his early preoccupation with the project of an ideal literal community in which to live.

We have now got so far on that the writer in the *Times Literary Supplement*, admonishing us to turn our minds on the problem of *using* the computer, actually ignores the problem: the word 'use' as he relies on it in making his admonition seem stern, solemn and cogently practical remains a mere word, empty of meaning and intention in regard to the essential problem that menaces mankind. The significance of his article lies in the innocent unawareness with which he exposes the emptiness. The worst kind of difficulty he recognizes as standing between us and the achieving of a satisfactory 'use'—such a use as will save humanity—is conservative and irrationally apprehensive resistance; resistance to what people must be made to see as benign change. They are afraid of losing their jobs; computers are (I quote) 'unambiguously linked to fears of down-grading and dismissal'. Again:

> Unfortunate resistance to change is equally strong among the professional educators, and modern teaching methods advance very slowly. Teachers fear that the increasing emphasis laid by educational theorists on teaching by remote control will lower their living standards by decreasing job-opportunities. What they fail to realize is that education is a growth area; it must be, if we are to survive as a nation.

'If we are to survive as a nation'—the irony of that! If this country is to survive, then we must make survival equivalent to the loss of all distinctive life, and we may hope to succeed. 'Growth' in 'education is a growth area' is not an insistence on organism; it doesn't, in fact, mean anything organic. The use of the word 're-structure' is profoundly significant: 'education' is to be re-structured, society is to be re-structured, life is to be re-structured. The sense that 'organic' means something is out of date; it has been left behind, and will hardly be found among our statesmen, the re-structurers they commission, or the mass of the enlightened who determine the effective re-forming mentality. The organic isn't congenial to statistical treatment, and therefore doesn't matter; it can be forgotten, and is.

Just as I had written that sentence my eye was caught by this, among my scrawled extracts from the *Times Literary Supplement* article:

> Finally, a computer, linked to some of the teaching aids listed above, can maintain an up-to-the-minute record of individual pupils' progress and draw the attention of the teacher to any falling away from pre-determined standards.

If anyone is inclined to see reassurance here, I can only say that I don't.

You will have recognized that it's not merely the professional subhumanities of computerial addiction that I've been sampling for you. The writer in the *Times Literary Supplement* is only drawing on the currency in which the prevailing 'advanced' ethos of our civilization expresses itself. Mr Fowler, Minister of State charged with supervising higher education, would find an armoury of practically minded particular suggestions and formulations in the essay—if he needed to go there. But he doesn't; he's supplied already—that's beyond all doubt. And there's no need to think of Mr Fowler's (or, indeed, Mr Christopher Price's) crusading enthusiasm as perhaps just a little proleptic. The sentence has been passed, the authoritative decree hangs over us: by 1980 the number of students at universities is to have been doubled—regardless of difficulties and objections (academics will, of course, raise them), if not of financial cost—certainly not: there are other ways besides expenditure of dealing with *that* problem. As Lord Annan, to whom the ruling ethos of enlightenment owes so much, has in the course of his own career taken his opportunities of intimating, dons must work harder—the reductive force of 'work' here being made brutally plain. They'll in any case, when they've learnt how to exploit and rely on it, have the computer and the computer-network to help them.

Dons must work harder. Last week, having had to call at Oxford on my way back to Cambridge from Bristol, I was accosted in a bookshop by a very old pupil of mine who is head of a foreign language department at another university. A scholar and critic, intensely interested in literature—and so in the intimately related literature of his own language—he is admirably qualified to be the best kind of liaison influence. He was depressed. He remarked that he was so overburdened with administration that he had no time to read and think, still less to write. He added: 'Now we're doubling our numbers', and intimated that this, for him, was the end: defeat.

It isn't only heads of departments who, when faced with the growing numbers and (an almost inevitable concomitant) with a growing proportion of students qualified to claim university places, but not qualified to profit by them, will find it impossible to keep themselves the equipped and charged university teachers implied in any serious conception of a university. When are they supposed to get their own reading and thinking—what one may properly refer to as their own creative work (for it should be that)—done? That such a question could have a bearing on a university teacher's fitness for his function is a possibility implicitly dismissed by the schemes promoted as democratic expansion. In short, the idea of education entertained by the planners is essentially the computerial essayist's.

You must forgive me if I say again at this point (I am so accustomed to misrepresentation) that I am not proposing to ban the computer, but emphasizing the problem of ensuring that the use of the computer shall be really a use—that it shall be used as truly a means in the service of adequately conceived human ends. More generally, I am not suggesting that we ought to halt the progress of science and technology, I am insisting that the more potently they accelerate their advance the more urgent does it become to inaugurate another, a different, sustained effort of collaborative human creativity which is concerned with perpetuating, strengthening and asserting, in response to change, a full human creativity—the continuous collaborative creativity that ensures significance, ends and values, and manifests itself as consciousness and profoundly human purpose.

Nor only does the essayist exhibit a representative blankness in face of the problem; he betrays—again representatively—an implicit *hostility* towards the human nature on which such an effort depends and which it expresses. He writes: 'If mankind bitterly resents the intrusion of machines into his social environment, he still more bitterly resents their intrusion into his physical body.' That kind of resistance plays, as a matter of fact, no great part in the writer's argument. Moreover, while the first of his suggestions seems to me unfounded—men don't, if there's

no noise and no stink, resent the presence of machines—
the second is hardly relevant to a discussion of the computer. And here lies the significance of the passage; it
becomes explicit in his next sentence: 'Once again it is an
instinctive rather than an intellectual reaction, and once
again, broadly speaking, it is an incorrect reaction.'

I will avow that the instinctive resentment at having to
have a machine *in* the body does *not* seem to me a bad one,
however incorrect it may be intellectually—so much the
worse for 'intellect' as the computerologist conceives it. Of
course, an instinctive reaction may be dangerous, or in-
convenient, but that is no reason for aspiring—as Mr
Gradgrind did—to eliminate the instincts. The computerologist's bent that way is manifest. If the aim could be
achieved, we should have handed over living to the computer—or (the same thing) eliminated the meaning of 'use'.
'All those subtle essences of humanity' that (to complete
appropriately the phrase of Dickens's referring, actually,
to Mr Gradgrind) defy the utmost cunning of computers
would have ceased to exist as trouble-makers. And I think
of a sentence of Lawrence's: 'What could possibly come
of such a people, a people in whom the living intuitive
faculty was dead as nails?'

A very strong, persistent and resourceful creative effort,
then, is desperately needed—a collaborative creativity to
complement that which has produced the sciences. I won't
offer to repeat the account of the university by which I
have tried to enforce my conviction that the university is
society's only conceivable organ for such an effort. Nor
will I offer a summary of how I conceive the function of
university 'English'. I have merely been attempting
to lay some emphases that, prompted by serious mis-
conceptions and misrepresentations that go on getting

publicity, I thought not out of place. I will close in the same vein. I don't, then, like the way in which the word 'literary' figures in comments on my position. To describe the 'one culture' of my insistence—'there is only *one* culture'—as a *literary* culture is to falsify my carefully defined intention. And, when you consider, what kind of limiting force has 'literary', even for the distinctive work of an English School? I won't revert to the more difficult considerations—I have, in this and that place, dealt with them fully and faithfully—involved in my aversion from talk about 'literary values'. But let me remind you that literary history plays a large and essential part in English studies, and suggest, with an illustrative reference or two, how little merely 'literary' literary history is. One of my undertakings at Bristol this term has been to discuss with some students the great change that was precipitated in civilization in the latter part of the seventeenth century. In such a discussion, properly pursued, the approach will be one belonging to an English School; but the discussion, the study, will not be properly pursued—I mean among English 'literary' students—without some serious dealings with various kinds of what I may call historian's history; with the history of science, with the history of philosophy, with the history of religion, with social and political and economic history, with the great relevant fact of France.

I can say no more here about the liaison function, the focal function, of an English School, but I hope the reference—a mere gesture of reminder—serves as a sufficient insistence that I don't by the 'one culture' mean what 'literary culture' in either Snow's or Huxley's intention implies. The presence of the diverse studies *together*, not without significant reciprocal influence, is necessary if the university is to be the centre of consciousness for the community it ought to be.

Such a university would be the creative centre of an educated public; a public representative and guarantor of the cultural continuity without which there can be no hope of checking the confident destructive follies of enlightened statesmen, intellectuals, bureaucrats, educational reformers, Provosts of King's College. So Lawrence's 'living intuitive faculty', informing and enforced by adequate knowledge of the changing world, would be kept alive and potent.

I can conceive no other way. You may dismiss the conception as Utopian and the proposed remedy as lacking simplicity and the immediate obviousness that compels. You can't reasonably call it 'literarism'.

And remember: no conception worth taking seriously can appear anything *but* bold and difficult, postulating as of its very nature it must, some robustness of considered faith; no serious prescription for the disease, which is desperate, can be simple.

As for the computerologist's prescription—our ruling team's official prescription—for ensuring our survival as a nation, its callous irony must surely appal: it amounts to saving by cheerfully destroying all that makes survival worth fighting for. It means expediting the process by which the country, not only of Shakespeare, Dickens, Lawrence and Blake, but also of Blake's *bêtes noires*, Newton, Locke and Johnson (I think of all these when I speak of the 'one culture'), becomes just a province of an American world. My saying this is hostility, not to Americans, but to essential Americanization—under which they themselves wilt.

To them we can say, tactfully but earnestly: 'You have

as much right to Shakespeare as we have, and Dickens was once yours as well as ours.'

Looking round at this beautiful university city, I have said to myself: 'Surely here the creative battle to maintain our living cultural heritage—a continuity of profoundly human creative life—must seem worth fighting; must be seen as a battle that *shall* not be lost.'

VI
Pluralism, Compassion and Social Hope

PLURALISM, COMPASSION AND
SOCIAL HOPE

I N our time it is very necessary to insist that the most
important words—important for those troubled about
the prospect that confronts humanity—are incapable of
definition. You can't by defining them fix and circum-
scribe their life—for in any vital use they will *live*, even
disconcertingly: there lies their importance for thought.
To say this is not to condone, or be tolerant of, irresponsibility: quite otherwise. The righteous and confident irresponsibility that is the mark of the 'humanist' intellectual,
as of the enlightened in general among whom his clichés
pass current, is to be seen in what he does with such
words. He reduces them to cliché-use and cliché-status.
His clichés have behind them clichés of attitude; expressions and inducers of flatteringly plausible non-thought,
they make thought, unless in the way of recoil, protest and
repudiation, impossible. The defence of humanity entails
their reclamation for genuine thought.

It sometimes seems to me that the central one, to re-
claim which would be to 'place' standard enlightenment
for the cock-a-hoop folly it is, callously lethal in its witless
inhumanity, is 'social'. 'The individual condition', we are
told, 'is tragic', but—'there is social hope.' I commented
on that characteristic proposition from the famous Rede
lecture in my own notorious (I may perhaps call it) reply.
I have, however, to confess that what seemed (and seems)
to me the unanswerableness of my comment hasn't had
the endorsement of universal recognition. Where, I asked,

is the postulated hope located—where *could* it be located
unless in individuals? I half thought I might get from
someone the reply that when people had identified themselves (as they ought) with the 'social hope' they would
cease to brood on the inevitability of personal death, and
of 'tragic' things such as bereavement. Though no such
reply actually came may way, the implication is very
general: in fact it gives the cliché-attitude of the enlightenment that Snow personifies an essential element of its
appeal.

The appeal is more potent, perhaps, when the implication stops short of full explicitness. For when you put it
in Lord Snow's way you are pretty obviously offering the
world the nearest thing the enlightened 'humanist' can
achieve to the religious, and so direct a challenge can
hardly fail to evoke at least a sub-perception that the substitute is a poor thing—a mere substitute, and bogus, in
fact. Is it possible to think of Lord Snow—or any other
exponent of enlightenment—as tragic? Who can attach
any tragic value to 'tragic' used in that way? The word
doesn't go with the progressivist ethos, the pretensions
and attractions of which involve the relegation of 'tragic'
to the cheapening journalistic misuse. As Snow, for
example, uses it, its stylistic efficacy is that it beckons up
an essential word, one of the key-words, in the basic
vocabulary of enlightenment: 'compassion'.

Essential, and endlessly produced; relied on and re-
sorted to in a routine way without compunction or embarrassment. Thus Lord Annan tells us that the twin-pillars
of his stance are 'pluralism and compassion': these impressive words give us the unshakeable faith that has kept
him fearless, upright and firm in the thickest of the
miasma generated by me. By 'pluralism' he means the

right to be incoherent and opportunist—or (shall we say?) decently supple; to be a trusted ally of politicians who don't doubt that Progress takes us—must be, resolutely and ruthlessly, *made* to take us—headlong towards the Comprehensive University, and yet to be found impressive when he assures us that above all things he values the Intellect, and first and last would (his master-aim) have the university student infused with respect for that.

'Pluralism', of course, is a self-defensive inspiration, prompted by a consciousness of exposure. It's 'compassion', the accompanying word, that Lord Annan counts on to release in the reader the convinced response, massive and sympathetic, that will communicate a kind of weight to the empty audacity (or felicity) of the *trouvaille*. Such reliance carries the endorsement of common usage: Lord Annan hadn't to think of it; it just presented itself—proposed itself as an inevitable and infallible recourse. I recall an experience I had recently at a university I was 'visiting'. Between two formal appointments I had to 'talk' with a roomful of students engaged in Social Studies. The questions put to me, in accordance with the expressed design, regarded my conception of what the university should be in a technologico-Benthamite civilization (my coinage was actually produced). All too soon a familiar type of politician-student took the floor; though his stated purpose was to ask me one particular question, he in fact went on and on, and accelerated with marked resolution whenever I attempted to answer the question he had begun with. Finally, seeing that otherwise nothing of the hour would be left, I firmly interposed and stopped him. He was very angry. 'Dr Leavis's tears,' he told the assembly, 'are crocodile tears.'

The observation was ludicrously gratuitous. My actual

165

offence (apart from stopping him) had been my declining to take up the theme of the suffering Mexican peasant when our explicit subject for the short hour was the university in industrial England. Having disallowed the standard *démarche*, 'compassion', I was being stigmatized in that automatic and oddly inconsequent way as disqualifyingly callous—patently incapable of the compassion that no decent person is without (a criticism that disposed of my views on universities).

Lord Annan's use of the tactic is less naïve and more sinister, but the tactic is the same: a virtuous refusal to consider the real problem—for which there are no easy solutions. But more sinister; for the student-intellectual, so young and so assured of his unprecedented maturity, hardly needs to ignore what stares him in the face; the clichés and assumptions that have formed him make seeing almost impossible. And enlightenment is easy: it confers confident superiority, and it costs nothing. Naturally then, if—pointing to this and that obvious fact or consideration —you force him to the brink of having to recognize that the problem is quite other than he has been brought up to assume, and immensely more difficult, he will be angry— as my Social Studies student was.

Anger of the same kind, though less forgivable, is what will come from the eminences of enlightenment, too, as *their* response—if they respond overtly at all—to disturbing criticism of their basic assumptions. Of course, they themselves are products of the civilization they find congenial (what other would have made them eminences?), but they bear a heavy responsibility. They are the primers, the reassurers and the prompters of the immense pervasive 'élite' of politicians, publicists, educationists and innocent do-gooders who take for granted their possession of unquestioned moral authority, and do certainly represent the 'culture' that explains why Vice-Chancellors and experienced academics concede essential ground so disastrously and so often let themselves seem to endorse the brash 'democratic' premises that student-politicians habitually invoke—invoke as axiomatic.

'Compassion', then, which goes so inseparably with 'social hope', is not the admirable thing it takes itself for. In spite of the pretension to sympathetic disinterestedness implicit in the word, the actuality is self-indulgence. I am not thinking primarily of the point Lawrence makes when he observes that the compassion of 'social hope' is self-pity. The emphasis I have immediately in view is that the ostensible grave call to imaginative responsibility actually serves as a licence for the opposite—for evasion and desperately reckless unrealism: that is, for irresponsibility.

That kind of contradiction, that basic equivocalness, reveals itself in many forms when one examines the ethos of enlightenment. Of course, I am no more inclined to doubt that there may be genuine magnanimity in the selectest progressivist circles than to dispute that there was genuine compassion at work in the conceiving of the Welfare State that embraces us all. But when I gather (from scathing printed letters in the *Times Literary Supplement* and elsewhere) that the fact that Lord Snow's friends stand by him, vouch for his magnanimity, and corroborate the judgment that I myself have none and in any case can't write, is to be taken as a damaging reply to my own careful argument, my comment is: 'Well, naturally they stand together'.—They stand together because they are anxious to preserve their standing as an élite, and they feel that, to that end, Lord Snow, whom they did so much to create as a major luminary, must be loyally stood by.

They are very consciously an élite—an élite of the progressivist intelligentsia: the characteristic 'pluralism' (to adopt Lord Annan's word) is given there. 'Élitism' is a bad word, a term of condemnation in the progressivist armoury; but nothing is stronger in them than the assumption that *they* are an élite. The consciousness of being one is so basic in them, so essential to their sense of themselves and their unquestionable importance, that it has something of the strength of unconsciousness. Thus an eminence of what may be called the *New Statesman* milieu, his intention trained sternly on the defenders of the grammar schools, can advocate in the most formal public circumstances a rigorous all-round imposition of the comprehensive principle and, when asked whether it is not true that he has entered his own son at Eton, reply that he thinks one should have one's children educated with the kind of people they will be mixing with in later life.

There is nothing unique about that once notorious illustration of pluralism. That is why it has been forgotten: there is a general sympathy among the progressivist intelligentsia, who aren't tempted to bear hard upon the chance failure of tact in a pursuit of social hope. The habit of unconcern for consistency or clear thinking appears in that curious word, 'under-privileged', which they use a great deal. It has, perhaps, its advantages: when, for instance, their magnanimity of compassion demonstrates itself, as it so often does, in their speaking up in an egalitarian vein for the 'underprivileged' they needn't feel the faintest quiver of moral discomfort. Yet the privilege implicitly claimed in the assumption of an élite standing is not the less privilege because it can be presented as responsibility: it couldn't be distributed all round on the principle of 'equality', since its nature is to be inherent in, or dependent on, special qualifications. It is a privilege of the enlightened to tell the vast constituency of the 'under-privileged', in whose name they offer to speak, what it really wants—to use, for example, their assumed representative status and authority to fix on the country the progressivist convention that it happily embraces the idea of becoming, or being made to become, a 'multi-racial society' with a 'multi-racial culture'.

The word 'élitism' is a product of ignorance, prejudice and unintelligence. It is a stupid word, but not for that the less effective in its progressivist-political use, appealing as it does to jealousy and kindred impulses and motives. It is stupid, and perniciously so, because there must always be 'élites', and, mobilizing and directing the ignorance, prejudice and unintelligence, it aims at destroying the only adequate control for 'élites' there could be. I put 'élite' as I spoke it, in inverted commas, by way of recognizing the fact that the word covers a diversity of things, so that when you bring together and compare the ends of the range of diversity you might be inclined to say that they hadn't much in common. There are scientist élites, air-pilot élites, *corps d'élite*, and social élites (the best people), and the underprivileged masses know that professional footballers and BBC announcers are élites. The word 'élitism' has its use (which is military) when it is a question of destroying the grammar schools or making the universities 'comprehensive'.

I have spoken of the progressivist intelligentsia as conscious of being an élite, but if it *is* one, it's not in the sense in which air-pilots, medical authorities, the higher Civil Service, and the British Olympic team are élites. I think it would be essentially misleading to apply the term at all to the real educated class we need, though the word 'élitism' is being used, and will in the name of 'democracy' go on being used, to destroy any possibility of its being developed. We who are determined not only to resist, but to defeat and reverse the destructive drive should face that truth in all its formidableness. But the admonition isn't necessary; and no one who shares the determination can fail to know that what, rather, has to be faced is the problem of replying convincingly to the question: 'But what, against such a drive, which is the drive of a whole civilization, *can* be done?'

I was thinking of that problem when, on another occasion, I remarked—and I have been questioned about the meaning of the remark—that when a Conservative statesman left the political field for the academic it gave me no warm glow of confidence that a clear and unequivocal voice would now be heard, desperately as that was needed, against the destroyers and the habit of malign righteousness. I was not implying a suspicion of anyone, or aiming at any particular target. What I had in mind was the well-known maxim: 'Politics is the art of the possible.' No one, I believe, who has moved for years a statesman among statesmen can fail to be subdued in his innermost being to the spirit of that maxim.

But *we* create possibility—we are committed to believing that, and must tell ourselves so, for the clear consciousness confirms, strengthens and emboldens the intuition it expresses. In the 'we' I have intimated an essential element of my answer—my answer to the despairing question: 'But what, against such a drive, *can* be done?' It is a shift from the discursive distance to the evoking of what we immediately know—to the appeal to that. The 'we', of course, is limiting; *this* is not the 'democratic' political process, the process of majority decision, in which the influence of enlightened progressivists plays so large a part. But minorities can matter: convinced, responsible and courageous testimony will tell, and the body of those who recognize what they know, and the force and bearing of what they know, will gather strength.

We come back here to Snow's solemn (and representative) cliché-game of playing 'individual' against 'social': 'The individual condition . . . is tragic', but 'there is social hope.' This is to empty the word 'social', to empty 'society', of everything that is not congenial to the technologico-Benthamite ethos and can't be reduced to statistical or organizational terms. What, one may ask, is left? Corrective suggestion isn't a simple matter, so complex, subtle and basic to life are the considerations involved. To challenge the crude antithesis, individual-social, is to demand genuine recognition for the fact, nature and significance of human creativity.

Blake's testimony is profoundly true: he lays such emphasis on art and the artist because the artist's developed creativity is the supreme manifestation of the creativity inherent in and inseparable from life—strictly inseparable, so that without it there is no perception. Except in the individual there is no creativity (any more than there is hope—whether conceived as 'social' or not). But the potently individual such as an artist is discovers, as he explores his most intimate experience, how inescapably social he is in his very individuality. The poet, for instance, didn't create the language without which he couldn't have begun to be a poet, and a language is more than an instrument of expression. He is—let the fusion of metaphors pass—a focal conduit of the life that is one, though it manifests itself only in the myriad individual beings, and his unique identity is not the less a unique identity because the discovery of what it is and means entails a profoundly inward participation in a cultural continuity—a continuous creative collaboration, something that must surely be called 'social'.

'Identity' is Blake's word. He uses it in relation to 'self-hood', its antithesis. The individual as 'selfhood' wills ego-tistically, from his own enclosed centre, and is implicitly intent on asserting possession. As creative identity the individual is the agent of life, and 'knows he does not belong to himself.' He serves something that is quite other than his selfhood, which is blind and blank to it.

To use the word 'social', then, as Lord Snow does, is to evacuate 'society' and leave it empty of life; it is to sterilize those words in the technologico-Benthamite spirit, and cut them off from all vibration, resonance or hint of the pregnant human truths and significances they should be felt to portend. He does this unconsciously, of course, merely exemplifying the habit and ethos of our civilization. How potent and stultifying that habit is comes out in Lord Todd's presidential address to the British Association—I quote from the abridged version that appears as I write these notes in *The Times* (September 3, 1970). Lord Todd is a distinguished chemist and the Master of a Cambridge college. He opens his address with this sentence:

> Although science has expanded enormously and with increasing speed during the past couple of centuries, it has had of itself little or no direct effect on society.

Mere students of English literature know better than that. Science, scientific method and scientific thought, science as represented by the Royal Society, passages from Spratt's history of which in Dryden's time are familiar to them, had a profound effect on non-specialist intellectual ideals, on the habits of assumption and valuation that marked the educated, on the conception of Nature and of the cosmos and Man's place in it, on standards of civilized conduct, on the prevailing notion of civilization, on architecture, on ethics, on religion, on the English language. They have to consider these things, and not primarily the 'dark Satanic mills' of the incipient Industrial Revolution, when they inquire into Blake's antagonism to the linked Newton and Locke (whom he associates with Pope and Dr Johnson). The explanation of Lord Todd's curious opening pronouncement comes in his second sentence:

> Nor could it have, since it is a cultural pursuit akin, indeed, to music and the arts.

The completeness with which Lord Todd is imprisoned in the naïveties of the intellectual world he inhabits reveals itself here with some felicity. He performs his monstrous reductive operation with the confident modesty of one who, bringing to bear the intellectual grasp of the trained mind, says what can't be disputed. What is left of 'society'? Lord Todd has eliminated in any case the essential creativity, the elimination being only emphasized by his innocent reference to 'music and the arts'. They are 'cultural pursuits', and 'cultural pursuits' are what it is desirable to add on to the material, factual and organizable reality of 'society'. They are what 'culture' is.

Lord Todd shows himself thus to be significantly representative in a way of which he has no suspicion: with a sense of austere responsibility he offers to tackle the great problem, and unconsciously excludes from among the data the essential human and social reality. For him it has never existed, and it doesn't exist now as a possibility, however remote. Nevertheless, it is the essential human donnée. There we have the desperate sickness of our civilization; the physician is the sickness in person.

'Culture', in these days of the 'debate about the two cultures', Ministers of Culture and the Arts, high-level international conferences about culture, and leaders in *The Times* about the 'pollution of culture', is one of those indispensable words whose use and behaviour have to be kept under observation. Lord Todd, having placed science among the 'cultural pursuits', goes on disarmingly:

> it seeks only to enlarge our understanding of the world in which we live, and the universe of which our world forms a tiny part.

This statement he obviously takes to be a very modest one. But what, we ask, does he mean by 'understanding' in the given context? Lord Todd is a genuine and distinguished scientist, but in what sense does he 'understand' the universe more—significantly more—than Lord Snow does? This is perhaps a rude-seeming question, but it needed to be asked—though I shan't attempt any answer. The point that demands to be made immediately regards the way in which Lord Todd moves from 'the world in which we live' to the world that 'forms a tiny part' of the universe. But does he move?—he is clearly unconscious of moving, and, in a sense, that settles the question: he doesn't. To conclude that, however, is to open one's eyes to the full, and frightening, import of the matter-of-fact unconsciousness with which he empties the life out of the verb 'live'. That is what technologico-Benthamism does.

The world we live in is not the world that forms a tiny part of the distinguished chemist's—the scientific—universe. It is a world, a reality, of human values and significances which is created and maintained by continuous collaborative human creativity. Without it there would have been no science. When the cultural continuity (that is, the collaborative creativity) begins, as now, to fail, mankind, even in an advanced democratic welfare state finds itself suffering from mysterious unsatisfactions, unhappinesses, cravings, addictions and disorders which the resources of technologico-Benthamite civilization prove impotent to appease or prevent. Science and technology (if there is no conclusive overt catastrophe) will continue to progress, and at an acceleration, for they have developed something in the nature of automatism. To say this is to recognize at the same time that their portentous advance involves sustained continuities of collaborative creativity, so that some members of various élites may derive for themselves something of that sense of significance—the sense that life is significant, or that they are in touch with significance somewhere—which mankind cannot do without. But that something certainly can't satisfy the profound need of the truly human being among them, for it can't satisfy the profound human need of anyone who is fully human. And for humanity at large, distinguished and undistinguished alike, it doesn't exist.

It would be absurd to allege that Lord Todd proposes to make it exist by imposing more compulsory science on the schools; absurd, if only because he has clearly, as a prescriber for our sick society, never harboured—never had to repel from his mind—any suspicion that the sickness could be of the kind I have pointed to, or any idea that there could be that kind of sickness. But if he were asked what he hoped would be achieved by enforcing on the schools a practical recognition that 'natural science is as much a branch of culture as literature, music and the arts' and that 'in school curricula it should be treated equally with the standard compulsory subjects like English, history and the rest', what would he answer? The answer he thinks it sufficient to give in his address to the British Association is this—he gives no other: 'Unless this is done we shall never have a scientifically conscious democracy.' If he were asked what he means by that, it would be found, I am convinced, that it had never occurred to him that an answer could be needed. He assumes innocently that the phrase, 'a scientifically conscious democracy' explains itself. No one could have assumed it, least of all an eminent scientist, who had, in any serious sense of the verb, *thought* about the issue. Lord Todd is simply exemplifying the malign potency of the cliché-world ('culture') that formed him and made him its distinguished representative. Not having had intellectual energy to spare for real thought in this field, and being wholly unpractised in it, he assumes the so-called 'debate between the two cultures' and lets his vocational bias commit him to the Snow side. Michael Yudkin, the then Cambridge biochemist who allowed me to print his critique of Snow's Rede lecture along with my *Two Cultures?—The Significance of C. P. Snow*, comments on the naïvety of Snow's 'theme':

There are, regrettably, dozens of cultures in Sir Charles's use of the term, even if the gap between the scientist and the non-scientist is probably the widest.

Yudkin's insistence there is on the fact that the different special sciences are so different that 'mutual failure of contact and comprehension' between their respective representatives is to a very formidable extent inevitable. This obviously has a bearing on the nature, force and meaning of Lord Todd's contention that 'natural science is just as much a branch of culture as literature, music and the arts', and on its relevance to his desire to impose a load of compulsory science-instruction, firmly and with no nonsense, on the schools. Anyone who wishes to strengthen his grasp of the grounds for deploring Lord Todd's offer to make the famous Snow cliché-brew intellectually respectable should read the brief and pregnant critique by Michael Yudkin I have referred to.[1] When Yudkin, a scientist himself, testifies to this effect he seems to me unanswerable:

> For the non-scientist, an understanding of science rests not on the acquisition of scientific knowledge, but on scientific habits of thought and method. No matter how many scientific subjects a child studies to 'O' level, no matter how many lectures on scientific method he attends, or how much he reads of Bacon or Descartes or Hume or Mill, he will never understand the nature of scientific procedure until he reads at least a Natural Science course for undergraduates.

Having observed earlier that there is no reason why scientists, from their side, should not 'bridge the gulf' if they wanted to ('But it could only be a one-way bridge'), Yudkin concludes that the Snow line, if enforced, would have for main effect a weakening of non-scientific education in the schools. He closes with this:

> There is a real danger that the problem of the 'two cultures' may cease to exist. There will be no building of a bridge across the gap, no appearance of modern Leonardos, no migration of scientists to literature. Instead there will be the atrophy of the traditional culture, and its gradual annexation by the scientific —annexation not of territory but of men. It may not be long before only a single culture remains.

Yudkin, for the purposes of his critique (which was a review), takes over Snow's 'culture' and 'the traditional culture' as Snow uses the word and the phrase. My way of putting what Yudkin with so little gratuitousness forebodes is that cultural continuity will have been destroyed,

---

[1] Contained along with my Richmond lecture in *Two Cultures?—The Significance of C. P. Snow* (Chatto & Windus).

which means that the essential collaborative creativity will have finally wilted. We shall no doubt still read articles on the sensible use of the computer, but in fact the computer will use us (why shouldn't it, seeing that it can even now write poems?). In the age when Man has achieved, as the journalists and Lord Todd (not alone among Heads of Houses) tell us, unlimited power over his environment, *humanitas* will have lost all power of control.

That is not what Lord Todd fears; that kind of proposition has probably never been brought to his attention, and there is every sign that, if it had, he would have dismissed it as meaningless. When he says that 'the root of our present problems is educational', his intention is determined by what he means by 'education'. The problems he has in view are, as he sees them, to be met by simple commonsense adjustments—largely, it would appear, computerizable. The solution is, firstly, to have more science in the compulsory school curriculum, and, secondly, when we are making our reformed provisions for higher education, to be careful that we don't turn out an excessive proportion of scientists and technologists to technicians:

> Whatever system is finally adopted I hope we will bear in mind that we need far more technicians than scientists and technologists. If we train too many of the latter many of them will have to follow the career of a technician, for which their training was not designed and which they will tend to regard as 'inferior'. The result will be a frustrated white-collar class, with all the dangers to society that such a class implies.

This, then, is the meaning that underlies Lord Todd's naïve demonstration of 'pluralism' or the art of having it both ways—naïve, and no doubt unconscious. I am thinking of the mentality that enables him, while asserting that science as yet has had 'little or no direct effect on society', to prescribe a stern equality with English of compulsory science-instruction on the plea that 'natural science is just as much a branch of culture as literature, music and the arts'. It would, I think, be a mistake on his part to retort that literature, music and the arts have themselves had no direct effect on society (and 'direct', as Lord Todd uses the word, means 'real'—or effect on what really constitutes society), if only because he has involved himself inescapably, and in Snow's spirit, in the 'debate about the two cultures', and mustn't risk damaging for use the indispensable word. But what in any case will ensure him a kind of adverse comment from which the classical Rede lecture was immune is his tactlessly tactical emphasis on another word, 'democracy'—'a scientifically conscious democracy'. Everyone knows, of course, that 'society' needs more technicians than technologists, but to come so near to avowing bluntly that the main 'social' function of the reformed democratic school, with its compulsory 'cultural' science, will be to produce future technicians won't win the unanimous applause of the enlightened.

I am pretty sure that by the time I deliver this lecture Lord Todd's address will have been treated to adverse comment of other kinds. Since, in fact, the Rede lecture won that near-unanimity of applause, the applause due to the immediately acceptable statement of plain good sense, the climate has changed; simple-minded repose on social hope has become less easy. The recent revolution of mood in America was bound of course to have some effect in this country. But the developments of technologico-Benthamite civilization here have themselves compelled a recognition of intransigent realities that disturb and might well dismay. Trade-union voices have explained the official impotence to prevent unofficial strikes by the insufferable boredom, and worse, of the factory floor. No one needs to be told why wage-inflation can't be controlled; the public capable of being troubled *is* troubled—the menace is so intelligible that it understands without having to think. How should men brought up in a world that lays the emphasis so incessantly on money and what money will buy—men without the idea of any possible satisfaction in their work, which for them involves little responsibility and has no human significance—*not* give their votes democratically and righteously for each new wage demand? And how should the warnings of the economists, who are in this matter easy to understand, have any effect on the movers and promoters? What effect has the risk of venereal disease on the practice of promiscuity, whether among children or legally adult pop-cultists? (and where's the line between those categories?).

It is in and of our desperate sickness that the worst consequences of progressivist enlightenment get very little recognition. If one followed up the last question by asking how the confident promoters of uninhibited sex, and of robustly enlightened (and appallingly ignorant) sex-instruction administered to young school-children, could have anticipated what their success would lead to, there would be a fairly wide sympathetic response, I think, to the ironical implication, but hardly in so far as this regarded the basic non-medicable, though profoundly vital, concomitants—the consequences for creative human life, which involves a subtle, maturely wise and infinitely delicate relation between male and female. Yet, as the general response to this summer's (1970) massive demonstration on the Isle of Wight shows, there *has* been a change; a change that should be for us a cue to intensify our resolution and do what, in the spirit of that, we see in front of us to do. The suggestion that enlightened reductivism, the vacuity of life in a technological world, a consequent sense that (even though there are plenty of jobs going for technicians) this civilization has little to offer one, violence, destructiveness, condoned irresponsibility in regard to sex, drug-addiction, 'student unrest'—that all these are intimately related wouldn't now be dismissed with the easy jeer of a very short time back.

We mustn't, however, expect a general recognition of the nature of the sickness—even though Lord Todd's idea of it, together with his idea of 'society', is seen to be ludicrously inadequate: we shall not very soon have leading articles, relevant and pregnant, on the spiritual philistinism and essential inhumanity—the anti-humanity—of our humanitarian civilization. Such phrases, with the perceptions they register, must count on being met with blankness; they simply won't be understood—that is the plight of civilized man. Yet the change means that a convinced appeal to a still responsive human sense of the profoundest human need is not merely quixotic and desperate, and therefore absurd. Such an appeal, which will be a creative effort sustained, convinced and resourceful, is in fact an expression of the truly practical spirit, for there is no other way of beginning really to tackle the real problem. I am thinking, of course, of what should be the truism that the university is society's response to its troubled sense of something profoundly wrong; it is the organ of the new kind of creative effort that is called for.

In saying 'new kind of creative effort' I point to my reasons for disliking Lord Todd's way—it's the ordinary way—of referring to the 'traditional university'. He means little positively by 'university', and there is no need to discuss any conception he may be supposed to have. It is 'traditional' that calls for comment; used as Lord Todd uses it, it performs a too familiar reductive office. It implies that you accept, inertly or subserviently, a 'heritage' handed down from the past, or, taking the proper attitude in an age in which the human situation is unprecedented and rapidly changing, you see that you must break free from the past, with its obsolete answers to human problems, and place your reliance, rationally, on the knowledge, insight and technology of the present—a present liberated from the past for an intense and deferential awareness of a future that is always superseding it. 'They have the future in their bones,' said Lord Snow of his scientific friends. His intention was plainly euphoric, but we can't help thinking of those unintended consequences of man's mastery which already manifest themselves in his bones. Lord Todd, of course, is professionally more interested in the chemical control of nature than in the potentialities of nuclear fission. My point, however, at the moment is that the Snow-Todd mode of thought, or non-thought, eliminates cultural continuity, excluding thus any more adequate idea of 'society' than that which belongs to Social Studies.

If cultural continuity were altogether dead and forgotten beyond recall, the undertaking to create a 'possibility' that could affect the politician's sense of his art would be much harder to embark on in a convinced practical spirit. But the accelerating advance that has brought civilization to its present straits has been almost incredibly swift. The young, whom politicians, enlightened clergy and innocent do-gooders (more and less innocent) unite to flatter and court, can't believe—unless, as some of them do, they come from educated homes—that things were ever different (except, of course, for the worse in ways that, prompted by Lord Snow or Professor Plumb, we all know of). But where there are educated homes cultural continuity is not forgotten—which is to say, not yet altogether dead. I made the same point with the remark that, though the mass-civilization congenial to technology was rapidly reducing and debilitating them, there were still in this country the makings of an educated public. The necessary creative effort won't have come out of nothing; the spirit of it will prove itself to be not gratuitous but realistic—I mean 'not Utopian' when I call it 'practical'. At the same time, when I describe the function of the now marked plurality of universities as being to create an educated public that politicians have to fear and respect, it is very far from a proposal to perpetuate and multiply Oxford and Cambridge as they were half a century ago. I myself have no impulse, or reason, to see a model in Cambridge as I have known it.

It is obviously out of the question for me to recapitulate now the various careful attempts I have made to convey just how a real university would be a 'creative centre of civilization'. What I had best do is to make briefly some points that relate to my emphasis on 'cultural continuity'. I have more than once invoked the fact of language to convey the nature of the life that Snow and the rest, with the confident ignorance that destroys, ignore when they talk of 'the traditional culture'—the life that has to be defended, fostered and perpetuated. I must do it yet again; the recourse is inescapable, its effectiveness being that language gives us so much more than a felicitous analogy. A language *is* a life, and life involves change that is continual renewal. A language has its life in use—use that, of its nature, is a creative human response to changing conditions, so that in a living language we have a manifestation of continuous collaborative creativity.

Of course, these phrases cover a range of possibilities. The greatest works of English literature represent a collaborative creativity of a completeness that has been forgotten. Shakespeare was able to leave the English language enriched, suppled and charged for all who were to come after him because of the actuality and potentiality of the language he started with—the language created by the English people in their daily lives and speech, which, by way of the Church, were in touch with a higher cultural heritage. He could be the great popular dramatist and, at the same time, our supreme intellectual poet. The last, the ultimate, great writer to enjoy the Shakespearian advantage was Dickens; the conditions and the very sense of the possibility have vanished. In the technological age a McLuhan can announce (at least according to the papers) that mass-media have turned the world into a village, thus seeming to demonstrate that he doesn't know what a village was—which would hardly be a paradox even in one brought up on the Canadian prairie. Even in Welsh or Fenland villages, and even in the recesses of Cecil Sharp's Appalachians, the people get their 'culture', their notion of civilized *mœurs*, and their sense of speech vitality, from the telly. I needn't develop that pregnant theme.

Yet—I speak (an important point for us) in England —we still have the English language, and a language (once again) is more than an instrument of expression; it registers the consequences of many generations of creative response to living: implicit valuations, interpretive constructions, ordering moulds and frames, basic assumptions. Having said this I recognize that I am faced by a complexity. The English language is possessed by many more people on the North American continent than here. When one expresses distaste for American influence on

English in this country, one is not thinking of distinctive American usages as such but of the characteristics that betray (or inculcate) the crude human attitudes of the civilization implicit in them. Cultural continuity has suffered worse, has been broken more completely, in America; there is a gap. When American academics write on English novels from Jane Austen to D. H. Lawrence, it is not the frustrating ignorance of English life, the English social structure, and the changing civilization of the century that most astonishes the English reader; it is the utter insensitiveness to those refinements of perception, distinction, valuation and interest which imply the collaboratively created human reality they depend on, and voided of which the novelist's theme becomes a mere opportunity for such gratuitousnesses of 'interpretation' as the critic's need to be original may prompt him (or her) to contrive.

But it is the problem, the situation and the possibility *here* I am concerned with. We can't restore the general day-to-day creativity that has vanished; we shall have no successor to Dickens. But we *have* Dickens, and we have the English literature that (a profoundly significant truth) Dickens himself had, and more—for there is the later development that includes Lawrence. There *is* English literature—so very much more than an aggregation or succession of works and authors. It reveals for the contemplation it challenges—in its organic interrelatedness reveals incomparably—the nature of a cultural continuity, being such a continuity itself. I mustn't now make any attempt to summarize what I have said on other occasions about the way in which English would be, in the real university, more than just another 'subject'. There's immediate point, however, in recalling that my emphasis falls on

'co-presence'—the co-presence of the diverse disciplines
and fields of knowledge; that I would have the opportunities for 'consultation' across the borders made the most
of; and that between such collaborative contacts and the
intellectual intercourse they would promote I would have
the reverse of any insistence on a dividing-line at which
one said: '*This* is *ad hoc* and in a "curricular" way formally
academic, and *that* is casual and informal.'

The point I had in mind is made when I emphasize that
I am not in the least tempted to think of English as the
evangelizing presence among lesser breeds who must be
taught the way to salvation. A strong and vital educated
public represents a living cultural continuity, and that
manifests itself, if it effectively does, in the responsiveness
of a charged human maturity to the problems our world
confronts us with. Obviously, in the generation and renewal of this at the university, the creative centre, specialist
disciplines and specialist branches of knowledge have their
indispensable positive parts to play.

But it will be asked, even if you got your educated
public, what could it do in the situation you point to? That
is the scepticism of human defeat which belongs to the
situation. If you don't believe that, except in the field of
technology and computer-controlled organization, creative
human effort can create possibility, then you don't believe
in the creative human effort, and that is to acquiesce in
hopeless defeat. I haven't suggested what solution will be
found for the appalling problems represented by the boredom of the shop-floor and the disinheritance of the masses
—I don't know. In the nature of the case, nobody knows,
so nobody can tell us. It is the nature of a sustained effort
of collaborative human creativity such as I have been talking about that it creates, and re-creates, its sense of possible

solutions, further problems, and remoter goals as it goes on: its perception of problems and goals changes. But unless there *is* the resolute fully human creative effort, and unless its continuance and development are provided for, nothing to the point will be done: technology and pluralism, even with compassion and the computer in close alliance, can't save humanity.

It is easy, however, to think of some very important consequences that would follow if we had the university and the educated class that a truly practical spirit would see as the attainable ends to be worked for. Distinguished chemists, when they became Masters of Cambridge colleges, would not be as completely unacquainted with intelligent ideas of education and society as Lord Todd's presidential address to the British Association shows him to be. They would not as a matter of course confirm their cruder promptings by intercourse with such authorities as the historian who wrote *The Death of the Past* (a work that Lord Todd appears to have read and been impressed by). One would hardly find a Professor of Modern English History assuring the world, in print, as Professor Plumb did, that Sir Charles Snow was historically right when, in his Rede lecture, he pronounced (breezily sweeping aside the sentimentalists):

> For, with singular unanimity, in any country where they have had the chance, the poor have walked off the land into the factories as fast as the factories could take them.

Against that set this from an actual witness in nineteenth-century England, Thomas Hardy (*Personal Writings*, edited by H. Orel, page 188):

> 'This process, which is described by the statistician as the tendency of the rural population to the large towns is really the tendency of water to flow uphill when forced.'

There is, readily adducible, a superabundance of evidence in support of Hardy.[1] I was on the point of adding, 'as every educated person knows'—but the Professor of Modern English History in the University of Cambridge, it appears, doesn't. Reinforcing his defence of Snow, and what Snow stands for, he tells the reader in a footnote (page 49) to *The Death of the Past*: 'Another refugee in a never-never land of the past is F. R. Leavis, whose picture of nineteenth-century England is totally unrealistic as it must be emotionally satisfying.' Professor Plumb would, if he had read him, hardly dare to call Hardy describing the past he intimately knew a 'refugee in a never-never land'. What Hardy describes is a positive civilization that made the poverty and hardship he also describes (his point depends on that) acceptable to the 'rural population' who had, under economic compulsion, to suffer, with the utmost reluctance, its loss. And I should like to know where Professor Plumb would send his readers if they wanted to study my 'picture of nineteenth-century England'. Actually, with a righteous contempt for delicate scruple, he is merely reiterating more elaborately his 'authoritative' assurance that the irresponsibility of Snow's I challenged is sound history. For Lord Snow's view of the past is as emotionally satisfying to Professor Plumb as it is to Lord Snow: it justifies that natural satisfaction with the present they both prosperously adorn which is expressed in their view of the future. *Parti pris*, that is, explains the pertinacity of their ignorance.

The beneficiary's natural satisfaction, the habit of righteousness, together with will that isn't purely benign —these are common manifestations of successful progressivist enlightenment. But such achievements of suc-

[1] See the note (page 193) at the end of this chapter.

cess at the level of what is thought of as the higher intellectual culture were made possible only by a prevailing cultural sickness—the sickness that manifests itself notably in what is called 'permissiveness', which has for concomitant a loss in the 'educated' class of fastidiousness; that is, of the finer habit of discrimination. Is anyone surprised at finding (say) that the editorial obituary in *The Times* is one (with photograph) of Jimi Hendrix, 'A key figure in the development of pop music'? And isn't it, on reflection, surprising that we shouldn't be surprised? If we needed it, the report on the front page of the same issue ('Festival pop star dies after party') reminds us of the nature of pop art:

> His style on stage was characterized by highly amplified, tortured sounds from his guitar, which he handled with strong sexual overtones, writhing and wriggling in apparent anguish.

This is the art, and this is the ethos, that enlightened clerics in our day enlist in the service of religion—I have seen, at Cambridge, the advertisements outside the University Church.

Such solicitude for the spiritual welfare of the young relates very closely—as other notices I have seen outside churches bring out—to the much indulged-in virtue of 'compassion'. The way in which this virtue imparts its righteousness to blind and comfortable irresponsibility makes it formidable. It is only three or four years ago that a leader in the *Guardian*, protesting against a proposed tightening of restrictions upon immigration, concluded by admonishing the reader to think of the damage such meanness would do our 'image' abroad, and the suffering that we, a rich nation, should be, so uncharacteristically, refusing to relieve. Such innocence rapidly ceases to be good journalism, for the obvious truth stands there too starkly to be long ignored: we might multiply our intake of immigrants by fifty and make no difference to the problem of over-population in the undeveloped countries—though we should be ensuring catastrophic transformations in our own. But of course the appeal to compassion continues to be worked, and with an unrealism that is certainly not less gross or less injurious.

In the spring I had occasion to walk on many successive days through Cambridge by a way that took me past half a dozen churches. Since my passage was in the early evening, the footpaths weren't crowded, and, though my preoccupation was very painfully absorbing, I didn't fail, when the official date had arrived, to see as I passed each church that it was Christian Aid Week. Outside all of them, including the University Church and the Congregational church, fixed to the railings, the legend that invited my attention ran: 'Hunger is an unnecessary evil.'

So complete is the triumph of enlightenment in our time that I have found explaining, or attempting to explain, why I should have felt a sick fall at the heart as I read, and took the significance of what I read, a delicate matter. Isn't compassion that makes people put their hand in their pocket a good thing, and isn't it a good thing that the churches should join in the drive to arouse it?—How does one answer?

'Hunger is an unnecessary evil': in adopting that as a text for spiritual authority's appeal to the world the churches are endorsing and sanctifying all the falsities of progressivist enlightenment. In the first place they are endorsing that disastrous falsification of the factual realities which is necessary to the progressivist religion-substitute. They are lending themselves to the required assumptions, ludicrous as they are, about the possibility of Western

'democracy'—or of efficient bureaucratic government—
in black Africa (or, for that matter, in India). How essential these assumptions are to progressivist morale, and how
strongly the progressivist ethos prevails, is shown by the
difficulty Professor Andreski had in getting his book, *The
African Predicament*, published. It is a story of publisher
after publisher rejecting it before he at last found one
willing to bring it out under the firm's imprint.

Andreski (Head of the Department of Sociology in the
University of Reading) makes it plain that when one
describes the system that prevails in black Africa as 'klep-
tocracy', and points to those to the enrichment of whom
the money given or lent in aid inevitably goes in very large
part, and to the way in which so much of the rest is mis-
applied, one is not righteously indicting Africa or Africans.
One is profiting by new light on the nature of cultural
continuity, and on the human-createdness of the human
world we live in. Nepotism isn't wrong in Africa—any
more than it is in India, and the conditions of a disinterested, or competent, Western bureaucracy don't exist.
Nor will idealist young volunteers have much success in
teaching the natives the American way of life—or, for
that matter, the British way of life (for Mr Edward Heath
found himself uttering that sinister phrase not long ago—
he meant no harm, but that only makes it a more disturbing symptom).

In any case, you won't by subscribing money to bestow
food, and the means of self-help, on the 'underdeveloped'
countries do anything towards making hunger an unnecessary evil. Professor Andreski establishes that con-
clusively—which is why he found it so difficult to get the
book published. I wish I could believe that it will be
found, and used, in every university department of Social

Studies, for the universities should, and could, be generating a public well-informed and intelligent enough to see
the significant relation between sanctified 'compassion'
and that conniving indulgence, the 'pop-service'.

Let me in concluding emphasize (for—shall I say?—
Professor Plumb's benefit) what I have *not* contended and
not offered and what my position is *not*. I have not said
anything against real sympathy and generosity or intelligent aid. I have not said, or implied, that we could do
without statesmen and politicians, or that they should
forget the maxim that 'politics is the art of the possible'.
I have not even faintly suggested that I could tell statesmen, politicians and the higher Civil Service how a solution of the urgent, enormous and infinitely complex problem our civilization confronts us with is to be achieved. I
have not, where the university is in question, been superior
about scientists, mathematicians, historians, philosophers
(I needn't attempt a comprehensive list), or left it to be
supposed that the point and profit of co-presence would,
as I intend it, be a tincture of literary cultivation conferred
on the other-than-literary professionals. The 'university as
a creative centre of civilization' doesn't mean that. Again,
I have not been William-Morrising, and I have proposed
no ideal condition of humanity to be found in any past.
I have referred to Dickens as a great Victorian artist who
has a profound significance for us, but no one can read
Dickens, least of all his greatest novels, without being
made vividly and disturbingly aware of the poverty, misery,
oppression and mismanagement that made Victorian England so very much other than a Utopia. The profound
significance depends essentially on that—as on the anti-
Utopist positive outlook Dickens's art conveys.

My argument is that, with all the human resources of

knowledge, experience and wisdom we can muster, we should really confront for what they really are the problems of the present. That means that the spirit of the given creative effort I want so much to serve is no more Utopist than it is nostalgic and reversionary—that, positively, it is by the appropriate criteria truly practical, profoundly realistic, and (as only light and energy out of the past could make it) profoundly of the present. The obvious enough (I hope) clinching emphasis at this point is a gesture towards what surrounds us. Here, bearing the name of the historic city, ancient second capital of England, is this convincing evidence of modern skill, modern and humane architectural intelligence, and modern resources, seeming, on its beautifully landscaped site, to grow in its modernity out of the old Hall, the old lakeside lawns and gardens and the old timbered grounds. It is easy to see that the architects have been guided by an idea that kept them in living touch with true and highly conscious academic foresight, and that the idea of the university as I have been insisting on it isn't merely mine. Where the external presence is so clear a promise of the justifying life and reality it is hardly possible to doubt their being generated, and prevailing; that is why I take pride in being allowed to feel still associated.

NOTE TO PAGE 188

(a) *The Origins of Modern English Society 1780-1880* by Harold Perkin (Routledge & Kegan Paul Ltd., 1969)

Perkin is uncritically enthusiastic about the Industrial Revolution in England as an undeniably good thing in enabling England to support a much denser population with an increasing power of consumption, 'rise in living standards', etc. Nevertheless, as a conscientious historian he incidentally (and without drawing the obvious conclusions) notes some facts that are significant for the argument against Snow's and Plumb's assertion that the English workers trooped off the land into factories gladly and of choice.

For example, Ch. IV section 2 : 'The revolution in organization naturally required a large-scale redistribution of people between occupations, industries, regions and communities'. [But] '. . . the population of almost all agricultural villages continued to rise until mid-century' [i.e. the 19th] '. . . the labourers bred faster than the available jobs, and wages stagnated or declined'. 'Their very reluctance to migrate, however, is evidence that they could not be driven to the factories and towns, but had to be coaxed.' 'The pull was not so much directly into the new factories . . . as into domestic weaving and ancillary labouring, transport, building, canal, and later railway, construction, and into heavy industry like ironfounding and coal-mining.' 'There can be no doubt, however, that it was higher wages which were decisive in recruiting workers, whether men, women or children, for the factories.' 'This is confirmed by the very reluctance of the workers to enter the mills.' 'That is why the early mill-owners were led to recruit the little parish apprentices from the Poor Law authorities, the only category of workers who were forced into the mills. . . .' 'Even higher wages were not enough in themselves.' '. . . the early entrepreneurs . . . had to stimulate the consumer appetites of their employees in order to make them feel higher wages worth working for.' 'All this is not to deny that there was a large element of compulsion in the worker's situation, however. Given the increased population, however caused, without increased opportunities for settlement on the land . . . then the choice before the individual worker was either to work or to starve.' He then shows that 'the fear of unemployment was effective. The fact also that so many handloom weavers and framework knitters in the 1830s and 1840s preferred to starve rather than accept the discipline of the factory shows that many who did accept felt themselves to be driven there as a last resort.'

'This brings us to the wider question of migration from the countryside to the towns.' 'Where did the new town-dwellers come from, and why did they go there? The well-attested answer . . . is that they came not by any important transfer of population from the south to the north but mainly from the adjacent countryside.' 'In general . . . the great majority of migrants travelled only a short distance, and the more rapid growth of the industrial areas was due less to a large-scale transfer of people than to their own greater fecundity.' 'The bulk of the surplus population of the southern counties was absorbed by southern towns, above all by London.' [He cites two other authorities in confirmation who wrote: 'The statistics suggest little in the way of a drift of population from the predominantly rural South to the industrial North.']

(b) *English Churchyard Memorials* by Frederick Burgess (Lutterworth Press, 1963)

'By the middle of the eighteenth century the repertoire of monumental types was wide, showing a good deal of ingenious regional variation, particularly in such a new memorial as the pedestal tomb which had some pretensions to being minor architecture. . . . The growing popularity of churchyard monuments, ranging from the humblest headstone to such edifices as this (often erected to minor gentry or retired city tradesmen), encouraged a wide employment of craftsmen both in villages and towns, the latter including men of some local importance, who were statuaries and master masons engaged in building, often supporting more than one workshop.

'This individually organized craft, along with the integrity of design that was its conspicuous feature, gradually disintegrated in the course of the next century, a victim of the complex changes in social conditions that are conventionally described as the Industrial Revolution. For instance, the effects of land enclosure crippled the class of yeoman farmer, for generations the mainstay of the country gravestone maker's employment, and the old cottage industries which had established some degree of craft fellowship within the villages were killed by factory competition, and this also led to shifts in population disturbing the family heritage of skill handed down from one generation to the next.

'In the less tangible realm of social relationships, the natural acceptance of an aristocracy of birth, intellect and taste, which to a large extent had ensured a permeation of culture throughout the different ranks of society, was beginning to dissolve into envy, doubt and unrest. In this respect the understanding between the aristocrat and the master mason, based on a tacit respect for each other's interdependent abilities, was disturbed by the intervention of the professional architect, which in effect deprived the dilettante of his role as artistic arbiter and the craftsman of his function as a decorative designer. . . . This new dominance by both architects and clergy not only reduced the independent status of the mason but, it is reasonable to assume, did much to undermine in him that confidence which is inseparable from pride in work.

'By the middle of the nineteenth century the altered structure of village economy could no longer support the same numbers of independent craftsmen, many of whom were either forced to become wandering journeymen or to seek employment in the towns, where monopoly forms were becoming established similar to those of building contractors today. The individual ownership of local quarries which had formerly enabled the mason to operate in self-supporting independence began to be replaced by company investment. With the opening of cemeteries, vested interest in foreign marble grew to such proportions as to make these imports competitive and eventually to cripple the native stone industry.

'The Victorian etiquette of piety, which encouraged commemoration as a social obligation, filling the green tracts of cemeteries with a multitude of monuments, not only led many town businesses to engage solely in memorial production, but large firms to standardize designs . . . imported ready-made from abroad. Admittedly this system of prefabrication for stock was nothing new; it had been customary in the past for masons to cut blanks during the winter months when weather prohibited building or repairs. In this case, however, both design and inscription were under the deliberate control of one man and possessed the integrity of personal expression, whereas in subdividing the activities of carver and letterer the possibility of the final product's becoming a work of art was diminished.'

(c) *Victorian Book Design and Colour Printing* by Ruari McLean (Faber & Faber Ltd., 1963)

'The Victorian period was one of rapid change, of fertile invention, and of enormous vitality, all of which is reflected in its books. It was a period of gradually lowering standards in the design of nearly all manufactures, but one in which that decline was vigorously protested against by a few; it was also a period of craftsmanship, at least in printing, which the present age would do well if it could emulate. If artists and designers were fumbling with their new tools, at least the pressmen had mastered theirs, and could print blocks and type on paper with a skill never before, and now rarely, seen. This skill in machining is at its most obvious and basic in black and white work: but this was also the period of the first commercial colour printing—perhaps the most beguiling and certainly the most characteristic feature of Victorian books. Working in most cases without photography, preparing all blocks and plates with their own hands and eyes, the Victorian colour printers produced some fantastic results; their contribution to the art of the book was unique. These slowly acquired skills no longer exist, as many of their most fragile books will never again be seen as they once were; but enough remains.'

VII
*Élites, Oligarchies and an
Educated Public*

ÉLITES, OLIGARCHIES AND
AN EDUCATED PUBLIC

IT is characteristic of our time, when a circulation of a million isn't big enough to maintain a newspaper, that one has to explain what one means by 'educated public'— since it is decidedly not the educated public in question here that the 'class' Sunday papers form and cater for. The university should be a creative centre of civilization—my explanation is conveyed in my account of what I mean by saying that. When I develop the proposition by enlarging on our need of a real and responsible educated public there is a shift to the plural: the constitutive function of the universities is to create such a public, and keep it vitally charged, conscious of its responsibility, and properly influential. The plurality belongs essentially to the idea; if one is not explicit about the interplay between the different centres the force of one's intention isn't fairly conveyed.

By way of evoking briefly the nature of the problem, or disease, to which the given emphasis on our need of an educated public is addressed I will quote from what, thirty years ago, I wrote in *Education and the University*:

> American conditions are the conditions of modern civilization, even if the 'drift' has gone further on the other side of the Atlantic than on this. On the one hand there is the enormous technical complexity of civilization, a complexity that could be dealt with only by an answering efficiency of co-ordination—a co-operative concentration of knowledge, understanding and will (and 'understanding' means not merely a grasp of

intricacies, but a perceptive wisdom about ends). On the other hand, the social and cultural disintegration that has accompanied the development of the inhumanly complex machinery is destroying what should have controlled the working. It is as if society, in so complicating and extending the machinery of organization, had incurred a progressive debility of consciousness and of the powers of co-ordination and control—lost intelligence, memory and moral purpose. . . . The inadequacy to their function of statesmen and labour-leaders is notorious, depressing and inevitable, and in our time only the very naïve have been able to be exhilarated by the hopes of revolutionaries. The complexities being what they are, the general drift has been technocratic, and the effective conception of the human ends to be served that accompanies a preoccupation with the smooth running of the machinery tends to be a drastically simplified one. The war, by providing imperious immediate ends and immediately all-sufficient motives, has produced a simplification that enables the machinery, now more tyrannically complex than ever before, to run with marvellous efficiency. The greater is the need for insisting on the nature of the problem that the simplification doesn't solve, and on the dangers that, when this war is over, will be left more menacing than before, though not necessarily more attended to.

Education and the University appeared in 1943, and the subsequent years have reinforced in a frightening way the diagnosis, the admonition and the foreboding. For a disease of that kind there is no remedy that can be made to sound simple, immediately convincing, and easy to apply. The only truly remedial process, that which entails an intelligent conception of the university, can't be recommended with the propagandist advantage of a simple prescription, but an account of it points to the one creative way—no other can lead to a human recovery. The distinction dismissed with such easy virtue by Mr Harold Wilson, that between a university and an Institute of Technology, comes in here: there is the necessary insistence on the co-presence of the different specialist disciplines and fields of study with the humane centre that a university English School should be. I have tried on different occasions to convey my sense of the delicacy of a proper insistence on that word 'centre'. I will avow now that, involved in the need to explain that the 'centre' will be a field of study existing as such by reason of a distinctive discipline of intelligence it cultivates—a discipline *sui generis* that is special though not specialist, there is a delicacy of another kind. For the inquiry, pushed home, as to how far any actual English School fulfilled this requirement would compel a recognition humiliating for us who care. But there is no defence of English as a university study that doesn't make the claim.

Of course, it is *one* emphasis: the more comprehensive claim would associate it with an insistence on the unsurpassable way in which English Literature *is* a literature, giving us centuries of cultural continuity. I mean to be making that point with a disarming disclaimer of immodest pretensions when I ask, 'But what is literary history?', and in reply point out how far, intelligently conceived, it is from being concerned with the merely *literary* (whatever that is, or could be). To lay, however, the emphasis on literary history and the centuries-long organic continuity of English literature when suggesting the possible profit of 'co-presence' for representatives of specialist disciplines and specialist fields of knowledge wouldn't be a good initial tactic—it might very well fail to seem convincing or convinced. But there are philosophers, historians, physicists, mathematicians and even sociologists who would react with pleasure to being corroborated in their private judgment that the culture of the Sunday papers is contemptible. Intercourse with such, especially if it led to some perception of the signs by which a real intellectual discipline manifests itself in one who has experienced it, would be a challenge to members of an English School to think of every step they could take by way of identifying their own field of study with a discipline of intelligence unmistakably distinctive and real—such identification being essential (though this suggestion will be very widely received as paradoxical rather than axiomatic).

My immediate point is that the discipline of literary studies—that discipline which makes intelligence inseparable from sensibility—is one without which there can be no adequate attention paid to the problems of our civilization. No one will suppose all the senior members of a university, or even a majority of them, to be capable of the kind of recognition I have in mind. But the humanely perceptive, actual and potential, are those that matter; they —those capable of a strong positiveness of humane conviction (*humanitas*)—may be a minority, but it isn't from majorities that the creative stimulus comes. A minority can change the spiritual climate. In the university in which an English School worthy of a good scientist's respect was livingly 'co-present' there would (essential to the English School itself) be a more embracing community of those qualified for mature human perception and judgment—qualified, that is, to collaborate in the formation of 'educated opinion'; and the total active presence would make the university the creative centre of civilization we need.

The verb I have just invoked once again prompts me now to associate with it an explicit emphasis: the collaboration that builds up the human world takes largely, as Blake, the great vindicator of human creativity, knew, the form of creative quarrelling. It is not unanimity that characterizes a real educated public, but the profound active knowledge that human nature and human need transcend the blind assumptions of technologico-Benthamism and that those assumptions are disastrous. In the creative community of collaboration that would form in the university the members would generate between them—to speak analogically in a way regarding which I have already, in various contexts, said what is relevant—a language that made creative quarrelling possible; that is, made possible the implicitly collaborative interplay of different bents and convictions. The 'language' would validate the assumption that we have enough common ground between us to make disagreement both intelligible and profitable.

It will, for instance, have (I hope) been noted with how limiting a generality I have touched on the religious aspect of human need: no one will have misinterpreted the purpose and spirit of the mode, or think that I imagine myself to have disposed of that matter, having paid it all the attention it calls for. But where, in the language of intercourse, the implicit common ground of assumption is not there and active, difference, disagreement and controversy will hardly be creative.

I have remarked on occasion—and the occasion often comes—that here, in our country, the realization of this idea of the university is not the Utopian dream it might be elsewhere; there are favouring conditions. That there *is* advantage gets a significant recognition—not the less significant by reason of the large unawareness it seems to carry with it—in an article in *The Times* (September 30, 1970) entitled 'Prospect for revolt in universities':

> the revolt has been subdued in Britain compared with, say, the United States, France or West Germany. The continuing

reality of the Oxbridge model in British university structures reduces their liability to be seen as sinister bureaucracies. Further, universities here have nothing like the close relationship with industrial and defence research as do those in the United States. And the state has not yet used police forces in campus disorders as has occurred in France, in Germany, the United States and Japan.

The most significant word in that passage—it gives us the paradoxical unawareness—is 'campus'. It is an American word that until recently no one, because it didn't apply to British universities, thought of applying to them: the advantage the writer points to is inseparable from that fact. 'Campus' brings with it by implication all the American conditions—which are so rapidly becoming established here; the rootlessness, the vacuity, the inhuman scale, the failure of organic cultural life, the anti-human reductivism that favours the American neo-imperialism of the computer. What the writer in *The Times* is noting is that something less empty and computerizable than the American non-idea of the university hasn't yet been altogether forgotten in this country. The British university as represented by Oxford and Cambridge was a distinct and strongly positive organic life, rooted in history. I have no temptation to idealize the ancient university as I have known it. But, as in an age of reductivist enlightenment I often find myself insisting, where there is—the product of human creativity—strong organic life there is always the imperative challenge to critical-creative conflict. Not only is perfection unattainable; perfection implies finality, and no one today is likely to forget that there is always change. What is important is to recognize fully the nature of the change, and to see that nothing essential is lost—nothing of the complex conditions of continued creativity; that is, to see what human disaster change will hurry us into if not humanly controlled.

The barbarity of reformist enlightenment is the deadly enemy that must be defeated—the civilized barbarity, complacent, self-indulgent and ignorant, that can see nothing to be quarrelled with in believing, or wanting to believe, that a computer can write a poem. In the interests of self-congratulation it simplifies social problems by eliminating the life; the complexities it reduces them to are mechanical, or treatable mechanically—it hates the organic, and its simplifications kill. Because one is grateful for Dickens and the past conditions that made him possible one is not to be taken as blind to the miseries, squalors and inhumanities his art records. What matters, the supreme significance, is the art and the general human creativity it represents, and it is a genocidal illusion to suppose that indifference to the art and the conditions of it, even though indifference presents itself as zeal for social reform, will improve the lot of humanity.

A university that is really one—this is an immediately relevant truth—will make it possible for the student (who won't be just a 'student') to feel he belongs to a complex collaborative community in which there are his own special human contexts to be found, and will make him, in his work and the informal human intercourse that supplements it and gives it life, more and more potently aware of the nature of high intellectual standards. These two conditions are essential, and all reforms that threaten them are immeasurably destructive—they threaten the life of the whole community we live in.

What I am voicing, of course, is not Mr Christopher Price's spirit of progressivist reform, or Lord Annan's. Even Lord Todd, who, with a distinguished scientist's concern for standards, testifies in his presidential address to the British Association that only a minority of each age group is capable of profiting by work at real university level—that 'such a group of élite exists', he says, 'must be clear to anyone who had been' (before the triumph of enlightenment) 'concerned with higher education, and it should, indeed must, be given the opportunity to develop its powers to the full'—even Lord Todd makes his innocent bow to what the ominous word portends when he speaks of the need to have a 'scientifically conscious *democracy*'. He makes his bow, but what he is concerned for is to maintain undiluted and unimpaired the necessarily small élite of scientists, and see that the democratically reformed schools ('democracy' for him being a politically realistic and disarming word) are devoted to keeping up a good supply of material for the higher-educational mills that are to turn out the needed virtually unlimited number —since technological civilization advances so rapidly—of trained technicians.

Some scientists emerge from the battle to maintain standards at the university more disturbed than Lord Todd seems to be. Thus Professor R. V. Jones, F.R.S., writing in a soberly factual way in the *Times Educational Supplement* (October 9, 1970), reports catastrophic defeat from Aberdeen, and suggests, as *The Times* gives it in its summary, 'that it might be realistic to send the few good honours students the university attracts, with those from other universities, to some "centre of excellence" '. In a technological civilization the need to maintain standards in the sciences is generally recognized, and scientists may hope with some confidence that the necessary 'undemocratic' recourses won't be dismissed as such, but be allowed to achieve an adequate solution of the problem. Such 'centres of excellence', however, won't be creative centres of civilization. And even if Oxford and Cambridge are included among the favoured locations, the process by which, as universities, Oxford and Cambridge become more and more like the 'mushrooms of mediocrity' the multiplication of which Professor Jones deplores will not be halted.

There is no evading the need to challenge with insistent explicitness the righteous confusions and betrayals that recommend themselves with the words 'democratic' and 'democracy'. There is the issue of 'participation', about which the very knowledgeable article in *The Times*, 'Prospects for revolt in the universities', I have referred to says, far from gratuitously, that it 'was clearly launched by the left with wider political problems in mind than the committees of universities'. Yes, the extra-academic political background has been apparent enough, and has had its young-lecturer agents in the academic foreground, even if they have there, for the most part, kept tactfully to the rear. But the authorities, who see plainly enough that they must take their responsibilities seriously, could not have made the imprudent and most questionable kinds of concession they have so often made but for the confused and confusing climate of ideas and assumptions they live in.

I remember reading in *The Times* a letter from a student-leader pointing out that the shows of 'participation' as yet conceded were derisory, since, on democratic principles (universally, or at least generally, accepted—this was the implication), the majority vote should decide, and in any university the students formed the vast majority. The confident naïvety of this letter was significant—as the editor who printed it obviously perceived.

No one but a student-leader could have committed himself to that solemn demonstration with such unmisgiving roundness. But that the 'axiom' he invokes is in the air as an implicit unquestionable axiom has, in ways disconcerting as well as amusing, been (I know) revealed at more than one university in the initiatory 'staff-student' contacts that 'participation' entails. And the fact that the 'axiom' lurks vaguely—lurks immune from explicit challenge—at the bottom of the experienced minds that most certainly *know* better debilitates the full conviction needed if the embattled righteousness of student politicians (backed by lecturer ward-bosses and demagogues) is to be properly met and dealt with. It is, then, necessary to get some clear and explicit thought current about 'democracy' and 'democratic'.

Of course, the 'left' that, as the writer in *The Times* says, launched student revolt 'with wider political problems in mind than the committees of universities' doesn't itself fail to practise an experienced realism about the methods and ethics of political influence and political action. Not that accomplished field-politicians are necessarily wholly cynical when they exploit, as they commonly do, the naïve propensities now inherent in the word 'democratic': it is partly, at any rate, a matter of automatism, and reinforces the point I made about the climate and its potency.

I remember a *tête-à-tête* I had with a 'shadow' Minister of Education. He began by raising an issue that involved the difficult and delicate problem of getting anything done towards the achieving of the real university, and raised it in such a way as to make it plain that the problem didn't exist for him. Faced with nothing I could offer to discuss, I merely said: 'What are you going to do about it when you're a government?' He paused for a moment; then, lifting a rhetorical fist, replied with impressive resolution as one disposing of the essential problem: 'We'll smash the oligarchy!' Reminded in this discouraging way of the truth that politicians, with inevitable consequences for their habits of thought, necessarily think first of winning the next election, I could only return, 'There are oligarchies everywhere', adding that there were even said to be oligarchies in *his* democratic party.

There *are*, of course, oligarchies everywhere—necessarily. We call them 'oligarchies' when we feel that an attitude of mistrust is in place—as it necessarily is. I am not being cynical or pessimistic; not, that is, meaning to suggest that the grosser mistrust is necessarily justified. But executive authority and power and the final processes of decision can't *but* be vested in a few. It doesn't follow that 'oligarchies' don't need to be kept aware that they are subject to criticism, check and control. Even good 'oligarchies' need that—as good 'oligarchies' don't often need to be rudely reminded. My politician's provocativeness lay not merely in his being so blatantly the politician, implying as he did that what he proposed to replace the smashed oligarchy by wouldn't *be* an oligarchy. It was that I had every reason for my conviction that it would be a far worse one, with disastrous consequences for the cause I had most at heart.

I must, I know, expect to see it stated as unquestioned fact that, on my own avowal, I don't believe in democracy. I don't know what 'belief' may imply, but I am not more opposed to democracy than Lord Annan or Mr Harold Wilson himself is. Democracy, as we all (or most of us) really believe in it, means the arrangements and the habits that save us from the plight of Russia, where a privileged and self-perpetuating bureaucracy is clamped down on the country, and knows that it can't be moved or intimidated. It means the accepted right and power of the country to decide by majority vote in free elections, not indefinitely postponable, that it has (or has not) had enough of the present government and would (or would not) like to see it replaced by one that has an alternative party-backing, representing a proclaimed and propaganded different policy and programme.

The referendum is alien to the spirit of British democracy. And Mr Harold Wilson, having (inevitably) demonstrated the point in practice, has once again, in face of a motion deploring his government's undutiful and disastrous disobedience, affirmed that it is decidedly not the duty of even a Labour government to act on the instructions of even a Labour Party Conference. The Conference might very well pass a motion prohibiting oligarchy, but that would, and could, make no difference; a government with Labour in power is still a government, and the Cabinet the Cabinet. And it is probable that among the less important members of the numerous total government there will be an aggrieved sense that the 'oligarchy' monopolizes the power; probable, too, that, in the Cabinet itself, a considerable proportion of the personnel will be known to murmur—among themselves, or talking confidentially to someone with (say) *New Statesman* connexions —that the P.M., together with his cronies B and C, habitually exercises, and obviously intends to go on exercising, the power of decision oligarchically.

The pejorative implication as a rule colours the word, as it did when my shadow Minister clenched his fist and said 'We'll smash the oligarchy!', the implication being that he would scorn to belong to one himself—whereas I knew he belonged to a powerful one, and aspired to belong to a powerfuller. Human nature manifests itself very commonly in that way.

'Élite', too, is used pejoratively, and—'élitism' as a politico-social vice or aberration being the theme in the background—slips readily into the prejudicial association with 'oligarchy'. The educated public, even when it is called the educated class (the class of the educated—I put it alternatively thus with post-Marxist enlightenment in view) couldn't possibly be called an oligarchy, and, for all the hazy incitement of 'élitism', it should be obviously absurd to call it 'an élite'. It isn't a class, either, in the politico-economic sense commonly attached to that word —the sense that, reducing 'social' in the Benthamite and post-Marxist way, defines 'class' in terms of the 'material-ist' interpretation of human history. The educated class or public, intelligently conceived, comprehends people of widely varied social position, economic self-interest and political standing—standing (that is) in relation to the possibilities of political influence. Its importance, in fact, is conditioned by its diversity of presumable bent and its lack of anything like ideological unity. When, as may happen, it is moved to indignation, protest and resistance by one of those casual threats to human life which charac-terize our age of accelerating progress, that response tells because it so clearly transcends sectional interest or bias.

Ordinarily, the educated class presents its vital unity as essentially a matter of diversities—diversities that make it the public without which there couldn't be the creative differences (rising into creative quarrelling) that maintain the livingness of cultural continuity. It *is*, in fact, the presence of the continuity, and *that* constitutes its unity.

I have slipped, I see, into the present tense: the edu-cated public that Matthew Arnold's ironies, castigations and admonitions assumed hasn't yet wholly disappeared; that is why, faced by the computerophils who are happy to believe that electronic devices will soon have all the human capacities we need preserve—and more, and will before long do our living for us, and the whole euphoric army of the enlightened they feel around and behind them, we can fight without a sense of utter futility for the creation of a conscious public of the educated that is equipped for its responsibility, convinced of it, and therefore influential. The human problem is complex and multiform: élites and oligarchies—and great men too—are necessary, but so is that which can check, control and use them, and, except as such a public, it can't exist—there is no other conceivable presence.

I haven't for a moment suggested that the achieving, the generating, of the creative effort postulated—and it must be a sustained one—will be easy. On the other hand, for those who recognize the fact, the nature and the importance for mankind of the essential human creativity, the recognition of formidable difficulties in the way won't be the signal for despair and inertness. I am not, then, tempted to leave unrecognized the discouraging conditions that might be supposed to have taxed my own resolution and faith most severely. I have laid much emphasis on the part to be played by the university, and on, as essential to that, the part of English. The experience of a life-time has made me profoundly aware of the peculiar hatred any intelligent conception of the importance of English may expect to encounter in a university English School. I discuss the problem in my Clark lectures, *English Literature in Our Time and the University*, pointing out that the difficulty lies in the antithesis to mathematics that, as an academic study, English presents. For the selectors of recruits to a Department of Mathematics the criteria are unequivocal; when they are arriving at their selection, their integrity and disinterestedness are seriously engaged, and they are able to defy the disastrously inappropriate promptings of human nature. No such proposition holds in relation to English; and mediocrities will naturally band to keep out the disturbing presence—which is that of the truly qualified, or, if by some chance it has got in, to keep it from being what it is qualified to be—'disturbing'.

This evident enough and easily verified truth is part of the general truth that creative effort starts from: there are no easy or evident solutions. For disinterested intelligence and the creative impulse that goes with it the urgency of the human crisis is both a challenge and an opportunity; and it is a further truth that *their* work is never conceived, and so never initiated, by majorities. Democracy (if 'democracy' is to be a good word) won't function unless the community has a strong educated nucleus. I have spoken of the universities as the creative centres of the educated public we need, and now, when the idea of 'university' has been disembarrassed of the 'divisive' associations of the word 'class' in its political use, is the moment to justify and enforce that emphasis by getting its significance manifested to the utmost in evident fact. To say this, of course, is not to do anything but honour the truth that the university, if it is to have roots, can have them only in the community—that it is society's necessary organ. And the plainer it becomes that the supporting and fostering public it depends on in its performance of its social function is a robust reality the better, for the function can't be performed without the evocation of hostilities and jealousies both insidious and, in brutal ways, formidable. That is the sickness of our technologico-Benthamite civilization.

For this, as any day we may expect to hear a presidential American voice proclaiming, is the century of the common man. And if the profounder needs of the common man don't find, to serve them, uncommonness more broadly based and more enduring than a Churchill or a de Gaulle the common man's future won't be anything to look forward to. There must, then, be an educated public that commands recognition as an impressive reality—as it will not do if it isn't confidently aware of its influence and of its responsibility. But a public that isn't conscious of itself as a public, that doesn't know that it exists as such, is hardly one. And this brings me back to my opening sentence: in an age in which a circulation of a million isn't large enough to maintain a newspaper it is hard to explain to anyone who needs to have it explained what the necessary educated public would be.

In any case, no attempt at the required kind of explanation is in place now. For my purpose, what I have said in the course of my argument is explanation enough. One needs, though, to give some thought to tactics, for a characteristic of the age we live in is distraction, or an unconsciousness that is inattention, and there are many who could be disturbed to uncomfortable recognitions by the comment on what otherwise, if noticed, would be dismissed with a casual shrug. I remember being answered by the Master of a college who was also the occupant of a distinguished professorial chair in a humane field of study, I having thrown out a commonplace (I thought) about an aspect of cultural decline: 'But there are more good books published every year now than ever before.' I didn't, I confess, know what to reply. If I had known at that time of the Professor of Modern English History who flatly endorses Snow's cheerful doctrine of the Industrial Revolution, I might have tried a reference to that on my humanist. I doubt, however, its availing anything, for I also recall that he was both surprised and ruffled by my critical judgment of Snow as a novelist, whose productions he obviously took for modern English literature, the manifestation of contemporary creativity.

This reminiscence should be enough to evoke with some vividness the difficulty to which I have referred. And it serves me as the cue for a note on an aspect of the total effort—one of concentration and life-renewal—by which the sickness has to be countered. The educated public we so disastrously lack is one that will be, for artists and the intelligent response they need, the effective presence of 'standards'. I put 'standards' in inverted commas as a disavowal of false suggestion: the indispensable public represents the ability to modify, in response to significant creation, the implicit criteria by which it judges. I am of course not assuming that, where expression in words is concerned, the 'significant' is confined to the creative work we call 'literature'. The educated public we need—and this emphasizes the importance of the point that it mustn't be thought of as a mere aggregation of individuals—will represent, for the creative writer and the critic (both of whom require such collaboration—for it amounts to that), a general lively awareness, or a readiness for it, of the significance of (say) Whitehead, Collingwood and Polanyi: I exemplify with a line of creative thought that is clearly of major significance for non-specialist intelligence and sensibility.

But it is not for nothing that, where the university is in question, I have ascribed a central function to the English School. Where there is not the public the literary critic implicitly postulates and appeals to as he endeavours to impart, and that is, to define for himself, new perceptions and valuations, such writers as those to whom I have referred have little effect outside the limits of small

P   217

specialist publics. What in fact we want is the kind of public implied, at any rate as an ideal, by the reviews of the last century, where literary criticism had its place among the diverse intellectual interests of a cultivated mind in its non-specialist capacity. But that phase of civilization is now remote; what we find ourselves asking is whether literary criticism as a living cultural manifestation can be restored—whether we may hope to have anything approaching a real contemporary performance of the critical function; anything like an effective play of critical intelligence on current literature and the changing literary climate.

That *Scrutiny* was run from a university is very much to the point; only from a university *could* it have been run, but the conditions of that kind of outlaw enterprise—they looked at the time largely like disadvantages, and very grave ones—have vanished for good. The changes, however, have not been wholly adverse to enterprises inspired and powered by the same convictions: the need for such convictions has come incomparably nearer to achieving wide recognition than it was forty years ago. I say the *need* for them, because until uneasiness and apprehension have associated themselves with positive ideas that favour creative effort they have not issued in what I mean by 'convictions'.

There is, then, for some group of the convinced that is suitably placed and commands a nucleus of collaborative abilities, a clear if tacit invitation, or challenge, to try whether, in launching a review, it can justify the hope of attracting the contributing connexion needed to keep it going. And, as a matter of fact, there is beyond doubt a good deal of talent that remains undeveloped and unexercised in relation to its most serious interests because there is no organ in which to publish. Certainly, if the promise of a high standard and a continued intellectual vitality were strongly conveyed in performance at the outset, the word would go round, and there would quickly be a significant manifestation of sympathy and will to support; and steady fulfilment of the promise would attach a growing public. That would be an immeasurable gain: the process by which uneasiness and apprehension become positive conviction would receive a sharp stimulus, and go forward, and the new influence would tell far beyond the limits of certifiable 'circulation'.

It is not my business now to develop this particular theme. What was necessary was to insist that there can't be the educated public that educationists, editors and politicians can be uncomfortably aware of while the currency of ideas, attitudes and valuation-tips controlled by the intellectual weeklies and the superior Sunday papers remains without effective challenge—that is, while there is no provision for current communication at a higher level. 'Challenge' may sound an unrealistic word, but I am insisting too that there is a higher realism, and that the 'convinced' (to use my shorthand), knowing this, won't rest at the inert recognition that a world in which advertising revenue plays so decisive a part, with cultural consequences we are all aware of, is hostile to their essential purposes: conviction means undiscouraged creativity. There will be little danger of their losing sight of what 'creative *centre*' implies, for there will from the beginning be powerful motives for discovering all the support and collaboration that can be enlisted outside the university. Indeed, they will have told themselves that the truly encouraging mark of success—improbable, perhaps, but certainly, in the spirit of their enterprise, not to be despaired of—

would be unmistakable and welcome effects of influence to be observed in the world where editors are in no danger of forgetting the importance of the advertising manager. Improbable, but not to be despaired of: suppose, for instance, that a newspaper with something like the standing of *The Times* could, to the characteristics and qualities that gave it its standing—the general informed responsibility, the appropriate manifestations of this on the editorial page, the columns of intelligently selected letters alongside the articles, the kind of 'readership' that made such a selection possible—be inspired to add a regular weekly page of reviewing that included intelligent, coterie-free and boldly disinterested critical attention to current 'literary' publishing, that would hardly injure the circulation of the paper, and would certainly reinforce its influence and standing and the general real indebtedness.

I don't think editors or proprietors would be easily persuaded to take this last proposition seriously, and I am not hinting at possible *démarches*. What I had in mind was to emphasize that the higher realism involves clear-sighted ambition, explicit in its bold matter-of-factness. If we are to have the educated public we need, then its necessary awareness of itself as an informed and influential community, the locus within the whole social body of human (or humane) responsibility, will not be adequately provided for by the university-based organ I have imagined.

I haven't been thinking of this last as confining itself to literary criticism, and shall hardly be supposed to be implying such a restriction where the cultural side of the imaginable ideal newspaper is concerned. My explicitness about the need for proper literary reviewing is to be explained by the considerations that make the English School central to the university. The educated public, where there *is* one, is the effective presence of cultural continuity; English Literature, I have emphasized, is a focal manifestation of the continuity, and a lively general awareness of that truth—a due realization of its significance—will be expressed in and fostered by a lively genuine performance of the function of criticism in and for our time. As things are, the potential intelligent public whose recognition the truly creative writer needs to feel he can count on gets no help; the occasional disinterested and real review, though it no doubt earns sporadic gratitude, doesn't at all tell against the total effect of concerted and conscienceless misguidance. In the BBC world and the weeklies (under which head fall the Sunday magazine-sections and the *Times Literary Supplement*) coterie reigns —in such conscious security that it feels no shame, or at any rate shows no sign of it. The case may be gross, but that, where the gross falsities are familiar and congenially directed, and, so, welcome to the coterie, is a recommendation. What offers itself as creative work would challenge recognition in vain if it were uncongenial to expectation and really, or at all profoundly, challenging—that is, disturbing to habitual complacencies and settled attitudes.

We have here a characteristic of the ethos that prevails in the literary world; it is the tacit resolution always to be on top; always—even when an acknowledged great writer is in question—to be able to look on him from a higher elevation. Thus Dickens is a great genius, but 'so soon as he thinks he is a child', and the intellectual who tells us with the assurance of representativeness that Dickens wouldn't have been able to read and understand Bentham tells us also that he was a Philistine. A critic who himself can read and understand Bentham and who clearly doesn't suppose himself to be a Philistine expresses, in saying these things, a consciousness of superiority to Dickens—who of course is all the same a genius. With this reductive habit commonly goes the standard ego-exalting enlightenment that expresses itself in an easy certitude that its advanced views are absolutely and self-evidently right. I say 'certitude' and not 'conviction' because the part played by the ego (and the background of coagulated egos) in this certitude is unmistakable. What 'conviction' means, as I use the word, draws its strength from something more authoritative than the ego.

Standard enlightenment is an essentially destructive process—a lethal reduction of a living cultural complexity. To the more respectable of the enlightened one may apply a phrase of Blake's: 'closed up in Moral Pride', their egotism disguised for them by what they feel as the unanimity of all who matter, they reduce complex human issues to terms that, if valid, would justify their easy certitude: they *know*, and any disturbing challenge is perverse, wilful and reprehensible. *Ecrasez l'infâme!* The disastrousness of the damage done in this way to that cultural heritage which represents a long creative continuity of human experience is implicitly and pregnantly conveyed by these two sentences from a philosopher whose thinking is *not* focused on my present theme:

We may say that when we learn to use language, or a probe, or a tool, and thus make ourselves aware of these things as we are of our body, we *interiorize* these things and *make ourselves dwell in them.* Such extensions of ourselves develop new faculties in us; our whole education operates in this way; each of us interiorizes our cultural heritage, he grows into a person seeing the world and experiencing life in terms of this outlook.

That comes from Michael Polanyi's essay, 'The Logic of Tacit Inference' (*Knowing and Being*, page 148), and I think the whole essay might be very profitably read in the context of this discussion; it will dispose of the suggestion that my approach, in some pejorative sense of the adjective, is too 'literary'.

The heritage to which Polanyi refers, the organic continuity of a mature culture, entails what amounts to an intuitive and tacit recognition of the complexity of life and of all challenging human issues. Such recognition, inexplicit as it mainly is, isn't spiritually inert; it means a capacity for perceiving—and to perceive is to respond (at least in feeling and thought) positively to—values and possibilities that for reductive enlightenment don't exist. Inevitably, the reductive habit is what Higher Education as conceived by the enlightened may be counted on to confirm, at any rate in vast numbers of the expanded university population—the population so rapidly expanded, and by 'Boyle's Law' (as *The Times* in its leader put it) doomed rapidly to double itself regardless of cost (except in money).

The absurdity of offering to isolate as *sui generis* and apart some realm of 'literary values' is intimated by an episode I recall from the 1920s—its significance has become more obvious with the lapse of time. An Indian (destined to become a Pakistani) who was working with me admired very much *A Passage to India* (then recent), and asked me whether I couldn't arrange for him to meet Forster. With the help of my ex-tutor, who had been a contemporary of Forster's and, as a loyal Kingsman, was proud of knowing him, the meeting was easily arranged —I was present. My Indian, having elicited from Forster that he had been in India, in sum, less than three years, said: 'It's remarkable, Mr Forster, how much you know, and how fairly you hold the scales between the Hindus

223

and the Moslems; but there's one party to whom you're *not* fair'. Forster, brought to a certain sharpness of attention by the unexpected, looked his question.—'The British.' 'Oh, one can't be fair to *them*', replied Forster, and with that the matter was dismissed.

Neither I nor my pupil liked that: we felt that he relied (and relaxed) too easily on his tolerant *sans gêne*—offered (it seemed to us) as a disarming and winning candour. I say 'tolerant', because tolerance was an aspect of the supremely humane wisdom he came to be credited with as the promoters of the cult succeeded in making him a great contemporary writer. But the implicit judgment my Indian friend intended was that Forster was exposing a grave defect of perception, human insight and essential valuation. I am not guessing: I recall another episode. The same Indian—it was at the Marble Arch—having spoken briefly to a metropolitan police-officer, said to me: 'It's a great country that produced *him*.' He had neither a sentimental tendency to idealize the British police-force, nor a sentimental devotion to the imperial motherland. He was thinking realistically of the problems that would face his own country when it had achieved independence—a goal to the attainment of which he *was* devoted.

He remarked to me that while he thought *A Passage to India* would do good in this country, in India it would do harm—by (he meant) encouraging the natural 'liberated' tendency to undervalue the immense creative work for India the British had done; had, by dint of the 'greatness' they represented, been able to do. This was more than forty years ago, and I think there is reason, now, for emphasizing the harm that the Forsterian ethos has done *here*. I notice that Lawrence (*Letters to Martin Secker*) writes in July 1924:

224

> Am reading *A Passage to India*. It's good, but makes one
> wish a bomb would fall and end everything. Life is more
> interesting in its undercurrents than in its obvious, and E.M.
> does see people and only people ad nauseam.

This, from Lawrence, is a drastic judgment; it amounts
to a major diagnostic constatation: the full and profound
creativity that draws from sources closed to the selfhood
is lacking in Forster, and with the lack goes a falseness in
the offered wisdom, which actually is a self-sufficient ex-
ternality of 'poise' taking itself for profundity, and sure
of applause. I think of the reply reputed to have been
made by Forster to Lawrence's 'Forster, you're dead':
'Perhaps I am'. Modesty, no doubt—or perhaps; but if so,
then modesty can be equivocal. And it is to be noted that,
a year later than the letter I have quoted from, Lawrence
writes to Secker (August 1925):

> *St. Mawr* a bit disappointing. The Bloomsbury highbrows
> hated it. Glad they did. Don't send any more of my books to
> Forster—done with him as with most people.

In fact, the key word for the diagnosis of Forster's case
is 'coterie'. As an undergraduate at King's he found
shelter, reassurance and coterie-fostering there (see *The
Longest Journey*), and when, so rapidly, Bloomsbury began
to assume importance in the illusory 'great world' of culture, his name became current as that of a coming great
writer. I heard of him at school, where there was a Kingsman on the staff, and acquired *The Celestial Omnibus*.
Later, after the war, but still well before *A Passage to
India*, I was lent his novels by my tutor, who also was a
Kingsman, and, with a proper concern to form the young,
loyally imparted to his acolytes in *statu pupillari* an
awed valuation of E. M. Forster. The coterie that had

established Forster in this paradoxical standing before he had anything plausible to show had ensured its life-long possession of him. His genuine creative impulse manifested itself in his recognizing the major creativity apparent in D. H. Lawrence and voicing the recognition (which required some courage), but the acclaimed success of *A Passage to India*, a work which, however distinguished, implicitly disavows the deep and total creative conviction without which there is no great work of art, is what he rested on.

He was content to enjoy the coterie-engendered reputation; talent and intelligence, that is, hadn't been able to free themselves from what was coterie *in* him—as can be seen in *A Passage to India*. It is not only that his ironic poise is essentially suspect in its equivocalness, there being so much that is patently coterie-habit in his humour and wit—for that is what its assurance is; one has to judge his wisdom and enlightenment cheap, meaning that he hasn't earned them by the personal exposure they imply—he hasn't paid the price; they haven't in them the implicitly claimed profundity and fulness of experience. ' "Pathos, piety, courage—they exist, but are identical, and so is filth. Everything exists, nothing has value." ' It might be retorted that this is dramatic. But it isn't; for Mrs Moore herself, the elderly woman of the Marabar cave, doesn't exist—except as a contrived opportunity for Edward Morgan Forster to make, without the embarrassment of avowed personal utterance, what seems to him an impressive *statement* of his genuine Bloomsbury lack of creative conviction.

It was impressive to Bloomsbury, which could recognize profundity in it without suffering disturbance, and found no difficulty in assimilating the cult of Forster to the cult of the more directly usable Lytton Strachey. The significance of this coterie ethos which has played so important a part in our cultural history is manifested in the barrenness of E. M. Forster's post-*A Passage to India* years, which were many—a barrenness that isn't in the least paradoxical. The point that has to be made is that his part in the history is, very largely, to have given the ethos its respectability and prestige, and to have enabled its cheapening and destructive vulgarity to pass for something else.

The point had to be made; it is essential to a due recognition of the nature and urgency of our need—the need to have the educated public that is the only effective presence there *could* be of cultural continuity. What must be thought of as characterizing such a public is not unanimity; one has rather, as I have said, to invoke the analogy of a language. There will be enough in common in the way of basic values and assumptions and deep implicit nisus to make differences that look extreme the concomitants, or necessary manifestations, of a total collaborative life. Thus the public in its totality will represent that strong living sense of complexities which is needed, above all in a time of rapid change, to ensure that the achievements, spiritual and humane—the essential creative achievements—of our civilization shall be permanent gains, conserved in the cultural heritage. It ought to be possible to deplore the easy and equivocal injustice of 'one can't be fair to *them*' without becoming suspect of hankerings after lost empire.

Let it be explicitly *said*, then, that those who are really, and without illusion, intent on saving the country are consciously, and necessarily, the vindicators and defenders of its greatness, and, as such, are not less essentially opposed to the spirit of those who wish they could believe that Mr Heath's government will make England great again than they are to the kind of enlightened wisdom personified in Bertrand Russell. The greatness gone, there is nothing left to save, and the loss is humanity's.

INDEX